Topics in Mining, Metallurgy and Materials Engineering

Series Editor

Carlos P. Bergmann

For further volumes:
http://www.springer.com/series/11054

Andréa Moura Bernardes
Marco Antônio Siqueira Rodrigues
Jane Zoppas Ferreira
Editors

Electrodialysis and Water Reuse

Novel Approaches

Springer

Editors
Andréa Moura Bernardes
Jane Zoppas Ferreira
PPGE3M - Programa de Pós Graduação em Engenharia de Minas, Metalúrgica e de Materiais
Universidade Federal do Rio Grande do Sul
Porto Alegre
Brazil

Marco Antônio Siqueira Rodrigues
Programa de Pós Graduação em Qualidade Ambiental
Universidade Feevale
Novo Hamburgo
Brazil

ISBN 978-3-662-51120-6 ISBN 978-3-642-40249-4 (eBook)
DOI 10.1007/978-3-642-40249-4
Springer Heidelberg New York Dordrecht London

Printed on acid-free paper

Springer is part of Springer Science+Business Media (www.springer.com)

To Prof. Dr. Adão Mautone, who planted the ideas that led to this book

Preface

The electrodialysis process principles are well known for over 70 years, but the large-scale industrial applications were only possible after the development of the multi-cell stack design and the production of efficient and stable ion-exchange membranes. Although the largest application of electrodialysis, the desalination of brackish water, has been around the world before reverse osmosis, the other applications like the treatment of certain industrial wastewaters have only recently been explored.

This book is an integrated presentation of electrodialysis with the first chapters covering in a very comprehensive manner the transport principles, the control parameters and membrane materials and characterization, and the last six chapters covering applications of treatment of wastewaters in industries that are water intensive users like oil refineries, leather, and metallurgy.

The book is an outstanding contribution to the advancement of the application of electrodialysis as a large-scale technology for the treatment of industrial waste waters.

Lisbon, Portugal, 22 June 2013

Maria Norberta de Pinho
Chemical Engineering Department
Instituto Superior Técnico

Contents

Chapter 1
Introduction

Jane Zoppas Ferreira

Water is known to be an essential resource for Humanity. People around the world are convinced about the need of providing new and improved means of maintaining a healthy environment. Despite its abundance on the planet, water tends to become scarce because of disordered consumption, which contaminates our reserves and make them unsuitable for use, either for human consumption, or for industrial use. The most varied industrial processes use lots of water directly. This water must be adequate for use. Countries without major natural resources treat seawater, in so-called demineralization processes, to produce drinking water for the population. All this water that is used for either domestic or industrial ends, must be returned to reservoirs as clean as possible because its contamination will make its reuse impossible. Industries are having difficulties in obtaining proper water because of the growing water scarcity caused by pollution. As a consequence, treatment costs are increasing, which limits productivity.

In the last few years, several wastewater treatment processes have been proposed that would enable water reuse, with special emphasis on waste minimization (e.g. process modification, material recovery) and hazardous waste treatment (e.g. from surface finishing industries, destruction of organic solvents).

One of these processes employs electrochemistry. This science is concerned with the way that electricity produces chemical changes and how chemical changes result in the production of electricity. Although there are many established applications, modern research has led to a great expansion in the possible uses of electrochemistry with exciting future developments. In the environmental area, electrochemistry has the potential to monitor or remove various polluting substances that affect the environment.

It is in this context that electrodialysis offers an electrochemical technique that removes ionic pollutants from an aqueous solution, producing two new solutions:

J. Z. Ferreira (✉)
Programa de Pós Graduação em Engenharia de Minas, Metalúrgica e de Materiais (PPGE3M), Universidade Federal do Rio Grande do Sul (UFRGS), Porto Alegre–RS, Brazil
e-mail: jane.zoppas@ufrgs.br

A. Moura Bernardes et al. (eds.), *Electrodialysis and Water Reuse*,
Topics in Mining, Metallurgy and Materials Engineering,
DOI: 10.1007/978-3-642-40249-4_1,

one concentrate of ions and another consisting of almost pure water. The first solution can be reintroduced to the industrial process and the water can be reused. Associated to another field of research, such as membrane technology, this technique can be powerful in the treatment of industrial effluents.

The use of electrodialysis is particularly significant because it approaches membrane technology as an advanced environmental technology that enables the development of clean treatment sequences for the recovery of water in industrial processes. The promotion of the use of the electrodialysis process, and the synthesis and characterization of membranes, will require a multidisciplinary effort in different fields, such as materials science, physics, chemistry, and sanitary engineering. Water reuse is a subject of great interest that involves the scientific and business communities, the government and society at large.

Based on the results obtained by our research group in the past decade, this book presents novel techniques to evaluate electrodialysis processes, to synthesize ionic membranes and characterize their properties, and to show the potential use of this membrane process in the treatment of effluents generated by many industrial sectors, such as refineries, leather industries, and electroplating processes. This book is aimed at students, researchers and engineers who are interested in membrane technologies for the rational utilization and reuse of water.

Chapter 2
General Aspects of Membrane Separation Processes

Andréa Moura Bernardes

Abstract This chapter focuses on the current challenges of water and wastewater treatment aiming reuse. Membrane separation processes are presented and electrodialysis is compared to pressure driven membrane processes, especially reverse osmosis.

2.1 Water and Wastewater Treatment for Reuse

The increasing demographic and industrial expansion observed in recent decades has resulted in a rising commitment of the rivers, lakes and reservoirs. Water use, treatment and reuse are still a worldwide challenge. The management of water resources should always address the multiple uses of water and the most efficient water management system is the one that applies a water treatment process that is compatible with the subsequent use of water. Appropriate quality parameters should be set for each specific use of water. The quality of water used for public supply is often not compatible with the characteristics required for industrial use. Furthermore, drinking water quality parameters may not be the most suitable for the use in processes that require demineralized water.

Water and wastewater treatment processes are chosen mainly based on the initial quality of the water, on the parameters established by regulations and on the proposed use. Nowadays, water for public supply and domestic or industrial wastewater are usually treated by physicochemical or biological processes, as presented in Fig. 2.1.

A. Moura Bernardes (✉)
Programa de Pós Graduação em Engenharia de Minas, Metalúrgica e de Materiais (PPGE3M), Universidade Federal do Rio Grande do Sul (UFRGS), Porto Alegre–RS, Brazil
e-mail: amb@ufrgs.br

A. Moura Bernardes et al. (eds.), *Electrodialysis and Water Reuse*,
Topics in Mining, Metallurgy and Materials Engineering,
DOI: 10.1007/978-3-642-40249-4_2,

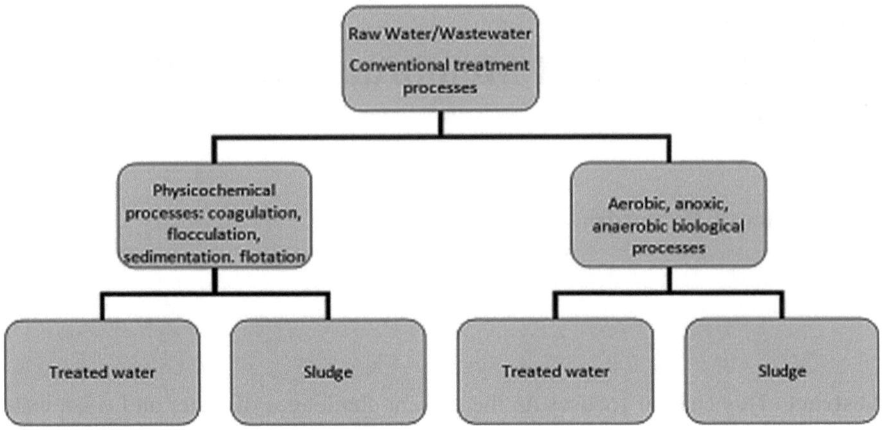

Fig. 2.1 Conventional water and wastewater treatment processes

These conventional treatment processes are not always suitable for potable uses, especially if the raw water contains nitrates, phosphates or other contaminants in small concentrations, which are not going to be removed by physicochemical or biological processes.

Considering the possibilities for the reuse of treated wastewater, the treated water generated after these processes can be reused, depending on the final characteristics, for such purposes as the irrigation of crops or landscapes, for the refill of aquifers and for non-potable urban uses, such as firefighting, flushing toilets, the washing of vehicles, streets and bus stops, etc... However, for industrial applications, such as cooling, boiler feed or processing water, as well as for drinking purposes, conventional treatment processes do not produce water that is of a high enough quality for reuse.

The growing demand for water has made the planned reuse of water a current topic of great importance. From this perspective, the treated effluents play a key role in the planning and sustainable management of water resources as a substitute for the use of water for industrial purposes, among others, contributing to the conservation of resources and adding an economic dimension to the planning of water resources. Reuse reduces the pressure on water sources because of the replacement of drinking water with lower quality water.

The membrane separation processes applied to water and wastewater treatment provide the use/reuse of these resources. These processes, however, have different properties and specific applicability. The ideal process would recover all the water, leaving behind only the salt. All current technologies produce a concentrate stream that must be discharged into the environment, or must undergo additional treatment to reduce the volume and remove the dissolved solids.

2.2 Membrane Separation Processes

The membrane technology has played an important role in developing more efficient and selective production with a reduced consumption of raw materials, energy and water and the minimization of wastewater and solid waste. Membrane processes have been introduced in industrial operations in order to treat the water, recycle process water and for the potential reuse and recovery of byproducts.

Different membrane separation processes are used in the treatment of water, sewer and industrial wastewater. Commercial membrane processes present different characteristics, as can be seen in Table 2.1.

Among the membrane processes presented in Table 2.1, pressure driven membranes (MF, UF, NF and RO) as well as electrodialysis (ED) are the ones applied to water and wastewater treatment.

In membrane separation processes driven by pressure, a pressure difference is applied across a membrane that can be of a Microfiltration (MF), Ultrafiltration (UF), Nanofiltration (NF) and Reverse Osmosis (RO) nature. The membrane acts as a semi-permeable barrier and may have different selectivities for different compounds.

Microfiltration is typically used for the removal of suspended solids or bacteria, using membranes with pore diameters ranging between 0.1 and 10 μ, and the transport mechanism is stereochemical exclusion, where molecules with a radius greater than the radius of the membrane pore are rejected.

Ultrafiltration is usually associated with the separation and concentration of macromolecules, using membranes with micropores of the order of 1–100 nm. The transport mechanisms that are generally employed in these kinds of membranes are molecular exclusion and diffusion. However, in many cases these mechanisms are associated to other phenomena, and the nature of the feed stream is very important, since the presence of colloidal material or the propensity to adsorb to the membrane leads to clogging and adsorption phenomena that alter the prevalent mechanisms of this type of operation (MF and UF). In certain cases, the productivity of membranes with quite different hydraulic permeabilities and selectivities can be shown to be similar, due to the clogging phenomena [1].

MF and UF technologies are typically used to treat water with high turbidity, to ensure the removal of viruses and bacteria from drinking water, or as a pretreatment in reverse osmosis systems [18].

Reverse osmosis is used to separate salts and small organic molecules from liquid streams, using membranes with dense active layers, where the preferred transport mechanisms are often attributed to solution/diffusion. Due to the high density of the active layer, operating pressures have to be much higher than those used in microfiltration and ultrafiltration.

The nanofiltration process is an intermediate separation process between reverse osmosis and ultrafiltration, commonly used in the separation of organic solutes with low molecular weight (200–1000 Da) and in the partial demineralization (essentially polyvalent salts) of liquid streams. The transport mechanisms that operate in these types of membranes are diffusion (as in reverse osmosis) and

Table 2.1 Membrane separation process characteristics [1 ,8, 17]

Process	Driving force	Typical separation mechanism	Retentate	Permeate
Microfiltration (MF)	ΔP (0.5–2 bar)	Sieve	Suspended solids, bacteria MW > 500000 Da (0.01 μm)	Water and dissolved solids
Ultrafiltration (UF)	ΔP (1–7 bar)	Sieve	Colloids, macromolecules MW > 2000 Da	Water, salts and compounds of low molecular weight
Nanofiltration (NF)	ΔP (5–25 bar)	Sieve + solution/ diffusion + exclusion	Molecules 500 Da < MW < 2000 Da	Water, salts and compounds of low molecular weight
Reverse osmosis (RO)	ΔP (15–80 bar)	solution/diffusion + exclusion	All soluble or suspended material	Water, solvent
Dialysis (D)	ΔC	Difusion	Molecules MW > 5000 Da	Ions and organic compounds of low molecular weight
Electrodialysis (ED)	ΔE	Ion exchange	Macromolecules and non ionic compounds	Ions

ΔP = Pressure difference
ΔC = Concentration difference
ΔE = Potential difference
MW = Molecular weight

molecular exclusion (as in ultrafiltration), but electrostatic interactions are also detected, which lead to selective removal of polyvalent ions [1].

Electrodialysis is a membrane separation process in which ions are transported through ion selective membranes from one solution to another under the influence of an electric field [2, 3, 20]. This transport generates two new solutions: one that is more diluted and one that is more concentrated than the original [4, 19].

2.3 Electrodialysis as an Alternative for Reverse Osmosis

Two membrane methods of water desalination, reverse osmosis (RO) and electrodialysis (ED) compete for a dominant position in a very large market, a competition that is continuously intensified by the increased shortage of water resources. Historically, electrodialysis had been developing faster than RO. In the 1960s, only electrodialysis was used in the industry. There were no RO units at all. The development of organic synthesis technologies has changed this situation, resulting in the manufacture of acetate and polyamide membranes, followed by free fiber high efficiency membranes. This development made RO began to dominate among the membrane methods. The construction simplicity of the units and the availability of membranes enabled even small companies to assemble the units and this promoted their entry in the market. In its two decades of operational experience, RO has demonstrated its strong properties, but the euphoria that accompanied its initial success has ebbed away and, today, the time has come for a matter-of-fact attitude towards both methods in order to choose the most optimal concept of membrane desalination [5]. Three factors need to be considered in evaluating the implementation of the two methods: the pre-treatment required for the feed water, in addition to the consumption of chemicals; the lifetime of membranes; and power consumption.

Membrane fouling is one of the most important factors that limit greater use of desalination membranes. Fouling occurs due to particulate matter, organic matter, microorganisms forming biofilms, and inorganic scaling [6]. Bacterial contamination problems have been reported as one of the main causes of membrane fouling in osmosis. This is caused by the characteristic of the membrane that serves as a barrier between the feed water and the product, which not only removes dissolved solids, but also bacteria, viruses, and insoluble substances.

ED only removes ions. Therefore, any bacteria, colloidal material, or silica present in the feed water stream will remain in the product stream. To minimize fouling and thus the need for the addition of chemical products, the polarity of the system can be reversed with electrodialysis reversal (EDR). By reversing the polarity (and the solution's flow direction) several times per hour, ions move in the opposite direction through the membranes, minimizing buildup. It is important to note that the EDR process does not directly filter the treatment stream through the membranes; contaminants are transferred out of the treatment stream and trapped by the membranes. This generally minimizes membrane fouling, decreasing pre-treatment requirements in comparison to RO [7].

In general, electrodialysis reversal (EDR) is more attractive for the desalination of brackish water and in cases where the removal of organic matter and microbial control are not important. EDR is also of interest in treating brackish waters where silica is an important limitation [6].

With respect to energy consumption, the literature indicates that the electrodialysis reversal process consumes less power when applied to waters with a total concentration of dissolved solids below 2,000 ppm. On the other hand, reverse osmosis is most advantageous when applied to waters with salinity greater than 4,000 ppm [8]. Other authors indicate that electrodialysis is more economical than osmosis in concentrations greater than 8,000 ppm when maintenance costs are taken into account [21] .

Several studies comparing desalination technologies with electrodialysis were performed on groundwater, surface water and effluents [9–15]. The main advantage of electrodialysis compared to reverse osmosis is that very little feed pretreatment is required, since membrane fouling and scaling is reduced to a minimum due to the reverse polarity operation. In addition, a much higher brine concentration can be achieved in electrodialysis when compared to reverse osmosis, since there are no osmotic pressure limitations. The chemical and mechanical stability of the ion-exchange membranes guarantees a long use life even in feed waters with aggressive and oxidizing components. Electrodialysis, however, has several severe technical and economic limitations. A major disadvantage, especially for the production of potable water, is the fact that only ions are removed, while uncharged components such as microorganisms or organic contaminants are not eliminated. Another disadvantage of electrodialysis is the relatively high energy consumption when solutions with high salt concentrations have to be processed. Likewise, the investment costs are prohibitively high when very low salt concentrations must be achieved in the diluate because of the low limiting current density, which requires a large membrane area. Thus, electrodialysis can only be applied cost effectively in water desalination for a certain range of salt concentrations of the feed water and quality of the product water. Outside this range of feed water composition and required product water quality, electrodialysis is not competitive when compared to other desalination processes [16].

Although water reclamation and reuse is practiced in many countries around the world, current levels of reuse constitute a small fraction of the total volume of municipal and industrial effluent generated [6]. Electrodialysis can be part of the solution in different applications, and this is going to be discussed in more detail in the following chapters.

References

1. Minhalma M (2001) Synthesis and optimization of processes for the recovery of industrial wastewaters with ultrafiltration and nanofiltration. PhD Thesis, IST, Technical University of Lisbon

2. Jamaluddin AKM, Kennedy MW, McManus D et al (1995) Salt extraction from hydrogen-sulfide scrubber solution using electrodialysis. AIChE J 41(5):1194–1203. doi:10.1002/aic.690410515
3. Solt GS (1971) Electrodialysis. In: Kuhn AT (ed) Industrial electrochemical processes. Elsevier, Amsterdam, pp 467–496
4. Birkett JD (1978) Electrodialysis. Berkowitz JB. Unit operations for treatment of hazardous industrial wastes. In: Noyes Data Co., New Jersey, pp 406–420
5. Pilat B (2001) Practice of water desalination by electrodialysis. Desalination 139(1–3):385–392. doi:10.1016/S0011-9164(01)00338-1
6. Escobar IC (2010) A summary of challenges still facing desalination and water reuse. In: Escobar IC; Schäfer AI Sustainable Water for the Future: Water Recycling versus Desalination, vol 2. Elsevier, Amsterdam, p 389–397
7. Seidel C, Gorman C, Darby JL et al (2011) An assessment of the state of nitrate treatment alternatives. Final Report. In: The American Water Works Association, Inorganic Contaminant Research and Inorganic Water Quality Joint Project Committees, California, p 136. http://smallwatersystems.ucdavis.edu/documents/ JACOBS-UCDavisNitrateTECReportFINAL062211.pdf. Accessed 10 Dec 2012
8. Baker RW (2004) Ion exchange membrane processes—electrodialysis. In: Membrane Technology and Applications. 2nd edn. John Wiley & Sons, Chichester
9. Lozier CJ, Smith G, Chapman JW et al (1992) Selection, design, and procurement of a demineralization system for a surface water treatment Plant. Desalination 88:3–31. doi:10.1016/0011-9164(92)80103-G
10. Swami MSR, Muruganandam L, Mohan V (1996) Recycle of treated refinery effluents using electrodialysis-A case study. Indian J Environ Prot 16(4):282–285
11. Van der Hoek JP, Rijnbende DO, Lokin CJA et al (1998) Electrodialysis as an alternative for reverse osmosis in an integrated membrane system. desalination 117(1–3):159–172. doi:10.1016/S0011-9164(98)00086-1
12. Post JW, Veerman J, Hamelers HVM et al (2007) Salinity-gradient power: evaluation of pressure-retarded osmosis and reverse electrodialysis. J Membr Sci 288:218–230. doi:10.1016/j.memsci.2006.11.018
13. Walha K, Amar RB, Firdaous L et al (2007) Brackish groundwater treatment by nanofiltration, reverse osmosis and electrodialysis in Tunisia: performance and cost comparison. Desalination 207(1–3):95–106. doi:10.1016/j.desal.2006.03.583
14. Macedônio F, Drioli E, Gusev AA et al (2012) Efficient technologies for worldwide clean water supply. Chem Eng Process 51:2–17. doi:10.1016/j.cep.2011.09.011
15. Li W, Krantz WB, Cornelissen ER et al (2013) A novel hybrid process of reverse electrodialysis and reverse osmosis for low energy seawater desalination and brine management. Appl Energy 104:592–602. doi:10.1016/j.apenergy.2012.11.064
16. Strathmann H (2010) Electrodialysis, a mature technology with a multitude of new applications. Desalination 264(3):268–288. doi:/10.1016/j.desal.2010.04.069
17. Habert CA, Borges CP, Nobrega R (2006) Processos de separação por membranas. E-papers, Rio de Janeiro, p 180
18. Antonia JM, Von Gottberg JM., Persechino AY (2012) Integrated membrane systems for water reuse. http://www.gewater.com. Acessed 29 Dec 2012
19. Genders JD, Weinberg NL (1992) Electrochemistry for a cleaner environment. The Electrosynthesis Company Inc., New York, pp 173–220
20. Rowe DR, Abdel-Magid IM (1995) Handbook of wastewater reclamation and reuse. CRC Press. Inc., Boca Raton, p 550
21. Ryabtsev AD, Kotsupalo NP, Titarenko VI et al (2001) Set-up involving electrodialysis for production of drinking-quality water from artesian waters with salt content Up to 8 Kg/ m^3 with productivity Ups to 1 m^3/h. Desalination 136:333–336. doi:10.1016/S0011-9164(01)00196-5

Chapter 3
General Aspects of Electrodialysis

Andréa Moura Bernardes, Marco A. S. Rodrigues and Jane Zoppas Ferreira

Abstract This chapter focuses on the general aspects of electrodialysis, presenting a historical background of the process and the technical principles. The electrodialysis stack design is introduced by discussing membranes, spacers and electrodes. Problems associated to fouling and scaling are presented. The phenomena known as fouling and scaling, which occur in membranes, can be avoided by applying the electrodialysis reversal. The basic concepts of electrodialysis reversal are presented. The efficiency of water and wastewater treatment by electrodialysis is evaluated through the calculation of percentage extraction and current efficiency. The total costs in electrodialysis are the sum of fixed costs associated with the amortization of the plant capital costs and the plant's operating costs, which are discussed at the end of this chapter.

3.1 Introduction

Although the large-scale industrial utilization of electrodialysis began about 20 years ago, the principle of the process has been known for about 100 years. The development of electro-membrane processes began in 1890 with the work of Ostwald, who studied the properties of semipermeable membranes and discovered

A. Moura Bernardes (✉) · J. Z. Ferreira
Programa de Pós Graduação em Engenharia de Minas, Metalúrgica e de Materiais (PPGE3M), Universidade Federal do Rio Grande do Sul (UFRGS), Porto Alegre–RS, Brazil
e-mail: amb@ufrgs.br

J. Z. Ferreira
e-mail: jane.zoppas@ufrgs.br

M. A. S. Rodrigues
Universidade FEEVALE, Novo Hamburgo–RS, Brazil
e-mail: marcor@feevale.br

A. Moura Bernardes et al. (eds.), *Electrodialysis and Water Reuse*,
Topics in Mining, Metallurgy and Materials Engineering,
DOI: 10.1007/978-3-642-40249-4_3,

that a membrane is impermeable for any electrolyte if it is impermeable either for its cation or its anion. To illustrate this, he postulated the existence of the so-called "membrane potential" at the boundary between the membrane and the solution as a consequence of the difference in concentration. In 1911, Donnan confirmed this postulate for the boundary of an ion exchange membrane and its surrounding solution. Simultaneously, he developed a mathematical equation describing the concentration equilibrium which resulted in the so-called "Donnan exclusion potential". The first basic studies related to ion-selective membranes were carried out in 1925 by Michaelis with the homogeneous, weak acid collodium membranes [18].

In 1940 Meyer and Strauss proposed an electrodialysis process in which anion-selective and cation selective membranes were arranged in alternating series to form many parallel solution compartments between two electrodes [18].

After the importance of the multicell stack arrangement for the economy of the electrodialysis was recognized, and with the development of synthesized ion exchange membrane by Juda and McRae of Ionics Inc. and Winger et al. at Rohm and Haas, electrodialysis rapidly became an industrial process for demineralizing and concentrating electrolyte solutions [18, 21].

The first commercial equipment based on Electrodialysis (ED) technology was developed in the 1950s to demineralize brackish water as an efficient and economic technique for the desalination of brackish water [19].

The main use envisaged for electrodialysis in the United States and Europe was the desalination of brackish water and seawater. In addition, Asahi Chemical Industries, Japan, initially focused on the concentration of sea water for salt production [16]. In the 1960s, the first salt production from seawater was realized by Asahi Co. with monovalent-ion-permselective membranes [10].

A significant step towards the efficient application of electrodialysis was the introduction of a new operating mode referred to as electrodialysis reversal by Ionics. In this operating mode, the flow streams and the polarity in an electrodialysis stack are reversed in certain time intervals and membrane fouling and scaling can be reduced to a minimum [17].

In the 1970s, a chemically stable cation exchange membrane based on sulfonated polytetrafluoroethylene was first developed by DuPont as Nafion, leading to a large scale use of this membrane in the chlor-alkali production industry and in energy storage or energy conversion systems (fuel cell), bringing about many new electrodialysis applications [10].

In April 1976, Dow Chemical Company announced that it would use the Asahi system in the United States. During the early years, the U.S. Department of Salt Water directly supported the research and development of electrodialysis for the production of potable water from brackish water. Similar programs were encouraged in Europe, Israel and South Africa, for the development of new membranes [16].

Apart from the initial desalination of brackish water [2, 3, 22], by the production of exchange membranes with better selectivity, lower electrical resistance, and improved thermal, chemical, and mechanical properties, other ion exchange membrane applications have recently gained a broader interest in the food, drug, and chemical process industry, including biotechnology and wastewater treatment

[10]. Electrodialysis is also being applied to the treatment of effluents containing various metals [4, 24]. Treatment of industrial wastewater by electrodialysis using NEOSEPTA membranes has contributed to pollution abatement and recovery of valuable minerals, such as the concentrating recovery of nickel from galvanization effluents. The treatment of copper-, zinc-, and tin galvanizing liquors and the recovery of noble metals are performed efficiently with this process [10].

3.2 Assembling an Electrodialysis System

The principle of electrodialysis is illustrated in Fig. 3.1, which shows a schematic diagram of a typical arrangement of electrodialysis cells, consisting of a series of anion- and cation exchange membranes arranged in an alternating pattern between an anode and a cathode to form individual cells. A cell consists of a volume with two adjacent membranes. An actual electrodialysis unit, which is referred to as a stack, may have a few hundreds of membranes [19]. In theory, a Faraday passing through a pair of membranes is capable of carrying one equivalent gram of electrolyte from one diluted compartment to a concentrated one. Thus, the insertion of n pairs of membranes will increase the yield of the process n times. From the electrical point of view, this system resembles a set of resistors in series. The resulting total electrical resistance includes contributions from the electrodes, membranes and solutions that flow between them. The limitation on the maximum number of membranes pairs, which can be mounted in a stack, is related to the increased total electrical resistance between the two electrodes. Ideally the resistance of the membranes must be extremely low, so that the main contribution to the total resistance is associated to the diluted solution produced between the membranes [5].

In a stack, the ion-selective membranes are alternately arranged in a filter press type assembly, in order to form channels between the membranes through which the solution to be treated circulates [14]. Within these channels spacers are placed, which have the function of causing a turbulent flow. The electrodes are positioned at the ends of the cell and are in contact with a rinse solution.

The driving force for the ion transport in the electrodialysis process is the applied electrical potential between the anode and cathode. When an electric field is applied to the electrodes, the anode is positively charged and the cathode is negatively charged. The applied electric field causes the migration of positive ions (cations) to the cathode and negative ions (anions) to the anode. During the migration process, anions pass through the anion exchange membranes, but are retained by the cation exchange membrane; a similar effect occurs with the cations. The overall result is an increase in the ion concentration in alternate compartments, while the other compartments become depleted simultaneously. The depleted solution is generally referred to as the diluate and the concentrated solution as the brine or the concentrate [19].

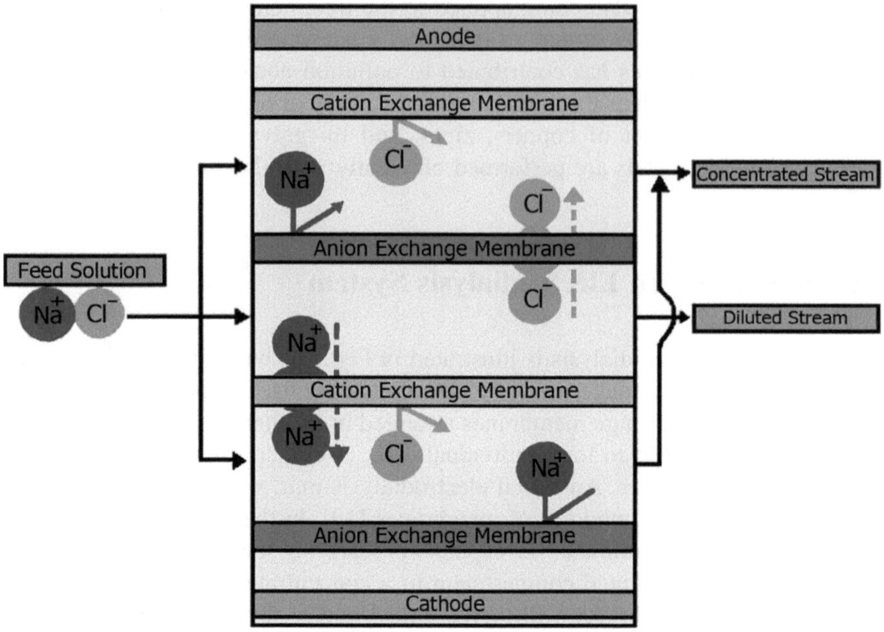

Fig. 3.1 Principle of the desalination process by Electrodialysis

An electrodialysis stack is a device composed of individual cells in alternating series with electrodes on both ends. The cell pair is a repeating unit in an electrodialysis stack. The electrodes containing cells are often rinsed with a separate solution that does not contain Cl^- ions to avoid chlorine formation [19].

A cell pair consists of the following [23]:

- Anion permeable membrane;
- Concentrate spacer;
- Cation permeable membrane;
- Dilute stream spacer.

Although many different components are necessary for the proper and efficient operation of an electrodialysis plant, such as the electrical power supply, pumps, and control and monitoring devices, the stack is the key element [19]. In each stack, different flows can be observed [23]:

- Source water (feed) flows parallel only through demineralizing compartments, whereas the concentrate stream flows parallel only through concentrating compartments.
- As feed water flows along the membranes, ions are electrically transferred through membranes from the demineralized stream to the concentrate stream.
- Flows from the two electrode compartments do not mix with other streams. A degasifier vents reaction gases from the electrode waste stream.

- Top and bottom plates are steel blocks that compress the membranes and spacers to prevent leakage inside the stack.

3.2.1 Membranes

Ion exchange membranes can be considered as ion exchange resins in film form [19]. The two types of ion exchange membranes used in electrodialysis are [23]:

- Cation transfer membranes, which are electrically conductive membranes that allow only positively charged ions to pass through. Commercial cation membranes generally consist of crosslinked polystyrene that has been sulfonated to produce $-SO_3H$ groups attached to the polymer, in water this group ionizes, producing a mobile counter ion (H^+) and a fixed charge ($-SO_3^-$).
- Anion transfer membranes, which are electrically conductive membranes that allow only negatively charged ions to pass through. Usually, the membrane matrix has fixed positive charges from quaternary ammonium groups ($-NR_3^+$), which repel positive ions.

Both types of membranes shows common properties: low electrical resistance, insoluble in aqueous solutions, semi-rigidity for ease of handling during stack assembly, resistant to change in pH from 1 to 10, operation in temperatures in excess of 46 °C, resistant to osmotic swelling, long life expectancies, resistant to fouling and hand washable [23].

Depending on the manufacturer, usually each membrane is 0.1 to 0.6 mm thick and is either homogeneous or heterogeneous, according to the way charge groups are connected to the matrix or their chemical structure. In the case of homogeneous membranes, charged groups are chemically bonded whereas heterogeneous membranes are physically mixed with the membrane matrix. Different manufacturers of ion exchange membranes operate in the market (Table 3.1). Each one offers membranes for specific applications, and they have different properties relating to their size, thickness, area resistance and composition [23]. Chapter 5 will discuss the properties and characteristics of ion selective membranes in more detail.

3.2.2 Spacers

The spaces between the membranes represent the flow paths of the demineralized and concentrated streams formed by plastic separators, which are called demineralized and concentrate water flow spacers, respectively. These spacers are made of polypropylene or low density polyethylene and are alternately positioned between membranes in the stack to create independent flow paths, so that all the

Table 3.1 Main manufacturers of ion exchange membranes [23]

Manufacturer/Reference	Country	Commercial brand
Asahi Chemical Industry Co.	Japan	Aciplex
Asahi Glass Col. Ltd	Japan	Selemion
DuPont Co.	USA	Nafion
FuMA-Tech GmbH	Germany	Fumasep
GE Water and Process	USA	AR, CR
LanXess Sybron Chemicals	Germany	Ionac
MEGA a.s.	Czech Republic	Ralex
PCA GmbH	Germany	PC
Tianwei Membrane Co.Ltd.	China	TWAED
Tokuyama Co-Astom	Japan	Neosepta

demineralized streams are manifolded together and all the concentrate streams are manifolded together too [23].

There are various spacer design concepts, but in practice only two different stack designs are used on a large scale. One is the so-called sheet-flow and the other the tortuous path-flow concept [19]. The main difference between the sheet flow and the tortuous path flow spacer is that in the sheet flow spacer the compartments are vertically arranged and the process path is relatively short. The flow velocity of the feed is between 2 and 4 cm/s and the pressure loss in the stack is correspondingly low, i.e. between 0.2 and 0.4 bars. In the tortuous path flow stack, the membrane spacers are horizontally arranged and have a long serpentine cut-out which defines a long narrow channel for the fluid path. The feed flow velocity in the stack is relatively high, i.e. between 6 and 12 cm/s, which provides a better control of concentration polarization and higher limiting current densities, but the pressure loss in the feed flow channels is quite high, i.e. between 1 and 2 bars [20].

3.2.3 Electrodes and Electrode Reactions

A metal electrode at each end of the membrane stack conducts DC into the stack. Because of the corrosive nature of the anode compartments, electrodes are usually made of titanium and plated with platinum. Its life span is dependent on the ionic composition of the source water and the amperage applied to the electrode. Large amounts of chlorides in the source water and high amperages reduce electrode life. Polarity reversal (as in EDR) also results in significantly shorter electrode lifetimes than for non reversing systems [23]. Typically, the lifetime of an electrode is two to three years and may be reconditioned.

The reactions that usually occur in the electrode compartments in electrodialysis are those that involve the formation of H_2 at the cathode:

$$H^+ + 2e^- \rightarrow H_2(\text{acid environment}) \tag{3.1}$$

$$2\ H_2O + 2e^- \rightarrow\ H_2 +\ 2OH^-(\text{alkaline environment}) \quad (3.2)$$

And at the anode, usually a reaction forming O_2 occurs, considering the use of inert materials as anode:

$$2\ H_2O \rightarrow O_2 +\ 4H^+ +\ 4e^-(\text{acid environment}) \quad (3.3)$$

$$4\ OH^- \rightarrow\ O_2 +\ 2\ H_2O + 4e^-(\text{alkaline environment}) \quad (3.4)$$

If the medium contains dissolved chloride ions a reaction forming Cl_2 can occur:

$$2\ Cl^- \rightarrow\ Cl_2 + 2e^- \quad (3.5)$$

The reactions occurring on the electrodes tend to acidify or to alkalinize electrode rinse solutions by the formation of H^+ or OH^-. The alkalization of the cathode rinse solution may lead to undesired precipitation of salts or hydroxides, such as $CaCO_3$ or $Mg(OH)_2$, if these substances are present. To minimize the effect of acidification or alkalinization of the rinse solutions, the same solution is used in the two electrodes, which is continuously mixed and recirculated in a closed loop [3].

3.3 Fouling and Scaling in Electrodialysis

The electrodialysis membranes are subject to deterioration by fouling and scaling [7, 12, 13].

Surface fouling and scaling are caused by deposition of suspended matter in a feeding solution on the membrane surfaces to form films. Small particles suspended in a feeding solution are usually removed using sand filtration, membrane filtration, coagulation sedimentation filtration, etc. However, extremely small particles pass through the filter, enter into an electrodialyzer and deposit on the membranes [21]. Suspended and colloidal matter, polyelectrolytes, organic anions, and multivalent salts near the saturation level can cause severe problems in electrodialysis due to precipitation on the membrane surfaces or by partial penetration into the membranes. Precipitation of suspended matter, silicates, and salts with low solubility, such as calcium carbonates or iron hydroxides, may occur within the actual flow channels, resulting in high hydrodynamic pressure loss and non-uniform flow distribution in the stack. Precipitation on the surfaces of the membranes also causes an increase of the electrical resistance of the stack and may lead to physical damage of the membranes [19].

Organic fouling of ion exchange membranes is one of the major problems in electrodialysis. It is caused by the precipitation of colloids on the membranes and, since most of the colloids present in natural water are negatively charged, it is almost always the anion exchange membranes which are affected. Organic anions such as humates, in particular, can precipitate on the anion exchange membranes as humic acid and cause a sharp increase in the electric resistance. Mechanical cleaning and treatment with dilute bases and acids can generally restore the original properties of the membranes. More severe is the poisoning of membranes by organic anions that are small enough to penetrate the membranes, but whose electromobility is so low that they practically remain inside the membrane, causing a drastic increase in the membrane resistance. Certain detergents are also the cause of this type of poisoning, which is very difficult to deal with and can best be avoided by a proper pretreatment of the feed solution [8, 19].

The physical parameters of the solute that have an influence on fouling are charge, hydrophobicity, molecular weight and solubility. The precipitation on the membrane is governed by the solubility of the solute, and adsorption is affected by electrostatic and hydrophobic interactions between the solute and the membrane surface. Furthermore, the size of the molecule also influences fouling. A large molecule moves more slowly in the membrane and blocks the path to other ions for a longer time compared to smaller molecules. The size of the molecule also affects the solubility and, therefore, the likelihood of precipitation increasing on the membrane [12]. The compound concentration and the current density are also important.

The literature [6] indicates that high molecular weight anions are responsible for the decrease in conductivity and permselectivity of the membrane. The permselectivity gradually decreases from 98 to 30 % when the number of carbon atoms in the molecule increases from 2 to 9, i.e. molecular weight of 59–171 Da. ED can be used in solutes with molecular weights up to 100 Da without significant changes in the properties of the membranes.

Another problem is the presence of microorganisms (biofouling); They can damage the membrane. The microorganisms can be destroyed during the ED operation by sterilization with ultraviolet and/or chlorine dosing [15].

Scaling is the formation of a crystalline precipitate of inorganic salt, usually at the cation side of the membrane, where the concentration of cations is high due to concentration polarization. Scaling affects the efficiency of electrodialysis significantly, and materials that precipitate on the membrane surface need to be removed with cleaning solutions. The cleaning frequency depends on the concentration of such materials in the feed solution [16].

Some components contained in waters that, to a lesser or greater extent, are responsible for the fouling and deposits in the ED membranes are [8]:

- Traces of heavy metals such as iron, manganese and copper;
- Dissolved gases such as O_2, CO_2 and H_2S;
- Organic and inorganic colloids;
- Particulate solids;
- Alkaline earth metals such as calcium, barium and strontium;
- Dissolved organic matter, natural or synthetic;
- Biological materials such as viruses, fungi, algae, bacteria.

3.4 Electrodialysis Reversal

The phenomena known as fouling and scaling that occur in membranes can be avoided by applying electrodialysis reversal. Various pre-treatment procedures, such as precipitation, flocculation or ion exchange and rinsing cycles, can be substantially reduced by a simple but very effective operating mode, which is referred to as electrodialysis reversal. In the electrodialysis reversal operating mode, which has been developed by Ionics Incorporated, the polarity of the electric field applied to the electrodialysis stack as the driving force for the transport of ions, is reversed in certain time intervals. Simultaneously, the flow streams are reversed, i.e. the diluate cell becomes the concentrate cell, and vice versa, with the result that matter that has been precipitated at the membrane surface will be redissolved and removed with the flow stream passing through the cell [1, 19]. The principle of the electrodialysis reversal operating mode is illustrated in Fig. 3.2, which shows a typical electrodialysis cell formed by a cation- and anion exchange membrane between two electrodes. If an electric field is applied to a feed solution containing, for instance, negatively charged particles or large organic anions, these components will migrate to the anion exchange membrane and will be deposited on its surface. If the polarity is reversed, the negatively charged components will now migrate away from the anion exchange membrane back into the bulk solution and the membrane properties are restored [19].

This procedure has been very effective not only for the removal of precipitated colloidal materials, but also for removing precipitated inorganic salts. In the practical application of electrodialysis in water desalination, the flow streams are reversed with the reversed polarity. Thus, the diluate stream becomes the concentrate stream and the concentrate a diluate stream. Generally, the flow streams and the polarity are changed every 30–60 min. The actual process of changing the

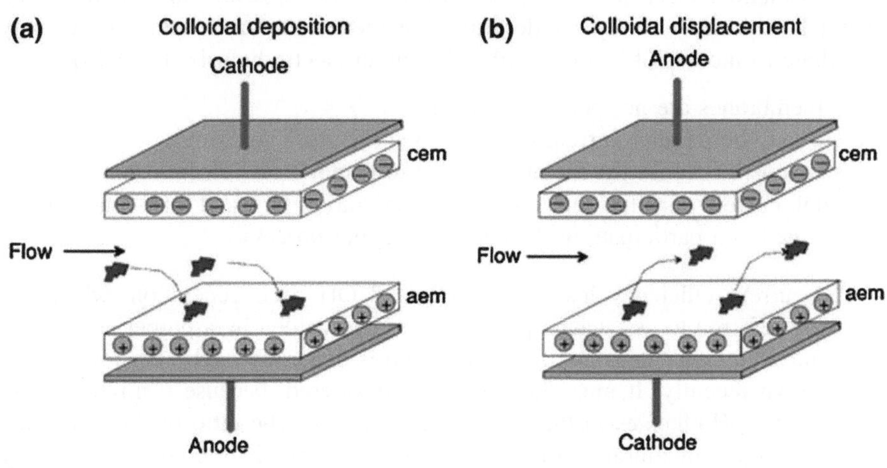

Fig. 3.2 Deposition (**a**) and removal (**b**) of colloidal matter in Electrodialysis reversal [19]

flow streams and the polarity takes only a few seconds. During this time the diluate composition does not meet the product specification and is disposed with the brine. Due to the polarity change, the productivity of an electrodialysis plant is reduced by ca 1–3 % [19].

3.5 Evaluation of the Process Efficiency

The efficiency of water and wastewater treatment by electrodialysis is evaluated by calculating the percent extraction and current efficiency.

3.5.1 Percent Extraction

The percent extraction of each ionic species in solution is calculated using the following expression (Eq. 3.6):

$$ep = \left(\frac{M_i^0 - M_i^t}{M_i^0}\right) \times 100 \tag{3.6}$$

where: M_i is the mass of species i and the superscripts 0 and t refer to the time of the beginning and end of the trial, respectively [9].

3.5.2 Current Efficiency

In any practical electrodialysis process, the current flowing through the stack cannot be utilized entirely for desalting the feed solution. Several factors may contribute to incomplete current utilization in an electrodialysis stack [19]:

- the membranes are not perfectly selective,
- there may be parallel current paths through the stack manifold,
- water transfer across the membranes due to osmosis and electroosmosis,
- at high current densities and low salt concentration H^+- and OH^--ions may be generated and participate in the current-carrying process.

The current efficiency loss due to H^+ and OH^- ion generation, which are generally referred to as "water splitting", can be avoided by a proper cell design and controlled concentration polarization, and is generally negligibly low, as has been shown recently. It must, however, be considered, because it might lead to considerable pH-changes in the electrolyte solutions. The ratio of the salt transported through a membrane from a feed to a concentrate is referred to as Faraday efficiency. This efficiency depends on the membranes used in the process as well as on current leakage through the manifold, which for its part depends on the system design and operating parameters. In addition, there is some transport of water through the membranes from the diluate to the concentrate solution due to osmosis and electroosmosis, which affects the process. The total current efficiency is therefore defined as the current required in practice to obtain a certain amount of product water of a given quality divided by the theoretically required current [19].

The current efficiency ec is defined as [25]:

$$ec = F \times \left(Vol_{0D} \times C_{0D} - Vol_{fD} \times C_{fD}\right) \Big/ n \times I \times \Delta t \qquad (3.7)$$

where: F is the Faraday constant, $Vol_{0\mathrm{D}}$ and Vol_{fD} are the initial and final volumes of the diluted solution, respectively, $C_{0\mathrm{D}}$ and C_{fD} are the initial and final concentrations of the electrolyte considered in dilute solution in equivalents. L^{-1}, n is the number of membrane pairs and I is the current passing through the system during the time interval Δt.

3.6 Electrodialysis Process Costs

The total costs in electrodialysis are the sum of the fixed costs associated with the amortization of the plant capital costs and the plant operating costs. Both the capital costs and the plant operating costs are proportional to the number of ions removed from a feed solution, i.e. the concentration difference between the feed and the product solution. They are also strongly affected, however, by the plant capacity and location and the composition of the feed water and the overall process design [20].

The capital costs in electrodialysis are mainly determined by the required membrane area for a given feed and product concentration. Other items such as pumps and process control equipment are considered as a fraction of the required membrane area. This fraction depends on the plant capacity. The total investment related costs for a given plant capacity depend not only on the required membrane area and the price of the membranes, but also on their lifespan under operating conditions, which in practical applications ranges from 5 to 8 years [20].

The operating costs are composed of labor, maintenance and energy costs. The labor and maintenance costs are directly proportional to the size of the plant and usually calculated as a certain percentage of the investment related costs. The energy required in an electrodialysis process is an additive of two terms: the electrical energy to transfer the ionic components from one solution through the membranes into another solution and the energy required to pump the solutions through the electrodialysis unit. Depending on various process parameters, particularly on the feed solution concentration, either one of the two terms may be dominating, and therefore determining for the overall energy costs. The energy consumption due to electrode reactions can generally be neglected, since more than 200 cell pairs are placed between the two electrodes in a modern electrodialysis stack.

The energy required for operating the process control devices can also be neglected in large industrial size plants [20].

The pumping energy depends on the solution viscosity, flow rate and recirculation of the concentrate, the need for disposal of the product and reject, the number of stages required and the efficiency of the equipment used. Energy consumption is affected by temperature. The electrical conductivity of saline solutions is quite temperature dependent, approximately 2–3 % per degree, because there is an increase on both the degree of ionization and mobility, reducing the electrical resistance of the solution and energy consumption. Another advantage is the reduced solution viscosity with increasing temperature, which results in lower pumping costs and a smaller width of the boundary layer at the surface of the membranes.

Some studies have shown an increase in the limit current density of 1 to 4 % for every degree of rise in temperature [11]. Other work has indicated that the energy required can be reduced by 60–70 % when the temperature increases from ambient to 70 °C [6]. As a rule, a 1 % reduction in energy consumption occurs for each increment of 0.5 °C for temperatures above 21 °C and a 1 % of increase in energy consumption for every equivalent decrease in temperature below 21 °C.

Among the disadvantages of elevated temperatures are the deterioration of the membranes and spacers associated with the degradation of the polymeric material and the increase of the tendency of precipitation that some compounds of low solubility present when the temperature increases.

The maximum operating temperature for projects of high temperature electrodialysis is between 50 and 70 °C, limited by polymer degradation. In conventional designs of ED/EDR the limit is 45 °C due to the loss of rigidity of the spacers, usually made of low density polyethylene.

The higher the current density, the lower the membrane area required for a given rate of desalting, decreasing the investment costs and the need to replace membranes. On the other hand, the energy cost is higher because the voltage increases in proportion to the current density.

References

1. Allison RP (1991) Surface and wastewater desalination by electrodialysis reversal. In: American water works association membrane technology conference, Orlando, March 1991
2. Andrés LJ, Riera FA, Alvarez R et al (1994) Separation of strong acids by electrodialysis with membranes selective to monovalent ions. An approach to modelling the process. Can J Chem Eng 72:848–853. doi:10.1002/cjce.5450720511
3. Applegate LE (1984) Membrane separation processes. Chem Eng 91(12):64–89
4. Asada K, Gerdes L, Kawahara T (1992) Electrodialysis of effluents from treatment of metallic surfaces. In: Proceedings of 79th AESF annual technological conference, Atlanta, June 1992
5. Audinos R (1983) Optimization of solution concentration by electrodialysis. Aplications to zinc sulfate solutions. Chem Eng Sci 38(3):431–439. doi:10.1016/0009-2509(83)80160-2

6. Awwa Research Foundation (1996) Lyonnaise des eaux-Dumez (Firm), South Africa water research commission electrodialysis. Water treatment membrane processes. McGraw-Hill, New York
7. Ayala EB, Pourcelly G, Bazinet L (2006) Nature identification and morphology characterization of cation-exchange membrane fouling during conventional electrodialysis. J Colloid Interface Sci 300:663–672. doi:10.1016/j.jcis.2006.04.035
8. Baker RW (2004) Ion exchange membrane processes- electrodialysis. Membrane technology and applications, 2nd edn. John Wiley & Sons, New York. doi: 10.1002/9781118359686.ch10
9. Gavach C, Lebon F, Pourcelly G et al (1992) Polarization Phenomena at the interfaces between an electrolyte solution and ion exchange membrane. J Electroanal Chem 336(1–2):171–194. doi:10.1016/0022-0728(92)80270-E
10. Kariduraganavar MY, Kittur AA, Kulkarni SS (2012) Ion exchange membranes: preparation, properties, and applications. In: Inamuddin Dr., Luqman M (eds) Ion exchange technology I:theory and materials. Springer, pp 233–276. doi: 10.1007/978-94-007-1700-8
11. Leitz FB, Accomazzo MA, McRae WA (1974) High temperature electrodialysis. Desalination 14:33–41. doi:10.1016/S0011-9164(00)80045-4
12. Lindstrand V, Jonsson AS, Sundstrom G (2000) Organic fouling of electrodialysis membranes with and without applied voltage. Desalination 130:73–84. doi:10.1016/S0011-9164(00)00075-8
13. Lindstrand V, Sundstrom G, Jonsson AS (2000) Fouling of electrodialysis membranes by organic substances. Desalination 128:91–102. doi:10.1016/S0011-9164(00)00026-6
14. Rautenbach R, Albrecht R (1989) Electrodialysis. In: Membrane processes. John Wiley & Sons, Aarau, pp 333–362
15. Scott K (1997) Handbook of industrial membranes, 1st edn. Elsevier Advanced Technology, Oxford, p 912
16. Strathmann H (1991) Electrodialysis state of the art, membranes—proceedings of India-EC workshop. Oxford & IBH, New Delhi, pp 25–69
17. Strathmann H (1994) Electrodialytic membrane processes and their practical application. In: Sequeira CAC (ed) Studies in environmental science—environmental oriented electrochemistry, vol 59. Elsevier pp 495–533. doi: 10.1016/S0166-1116(08)70563-6
18. Strathmann H (1995) Electrodialysis and related processes. In: Noble RD, Stern SA (eds) Membrane separations technology: principles and applications. Elsevier Science, pp 213–281. doi: 10.1016/S0927-5193(06)80008-2
19. Strathmann H (2004) Ion exchange membrane separation processes, vol 9. Elsevier, Amsterdam, pp 1–348
20. Strathmann H (2010) Electrodialysis, a mature technology with a multitude of new applications. Desalination 264:268–288. doi:10.1016/j.desal.2010.04.069
21. Tanaka Y. (2007) Ion exchange membranes: fundamentals and applications, vol 12. Elsevier, Amsterdam, pp 1–531
22. Thampy SK, Narayanan PK, Chauhan DK et al (1995) Concentration of sodium sulfate from pickle liquor of tannery effluent by electrodialysis. Sep Sci Technol 30(19):3715–3722. doi:10.1080/01496399508014154
23. Valero-Cervera F, Barceló A, Arbós-Sans R (2011) Electrodialysis technology: theory and applications. In: Schorr M (ed) Desalination, trends and technologies. InTech, pp 3–20. doi: 10.5772/14297. Available from: http://www.intechopen.com/books/desalination-trends-and-technologies/electrodialysis-technology-theory-and-applications. Accessed 05 Nov 2012
24. Walsh FC, Pletcher D (1993) Industrial electrochemistry, 2nd edn. Chapman & Hall, London, pp 331–450
25. Wisniewska G, Winnicki T (1991) Electrodialytic desalination of effluents from zinc-coatings processes: removal of ZnH^{+} and Cl^{-} ions from model solutions. In: Proceedings of the 12th international symposium on desalination and water reuse, Malta, 15–18 April 1991, pp163–176

Chapter 4
Electrodialysis Control Parameters

Luciano Marder and Valentin Pérez Herranz

Abstract Electrical conductivity, pH and concentration of the electrolyte, electric current or applied potential, are the basic control parameters of an electrodialysis system. However, to ensure the technical feasibility and an efficient use of electrodialysis processes, it is essential to understand the electrochemical behavior of ion exchange membranes, its stability, conductivity and selectivity. It is also important to observe some specific conditions of electrochemical processes, such as the occurrence of concentration polarization and limiting current density. Therefore, the main objective of the present chapter is to give a brief summary of the different control parameters that are important to decide which membrane is the most appropriate and to establish the maximum current density that can be used for a given application.

4.1 Membrane Permselectivity and Transport Number

The term permselectivity is related to the ability of an ion exchange membrane being simultaneously permeable to counter-ions and impermeable to co-ions. It represents, therefore, one of the most important properties of an ion selective membrane for a given application and it is one of the main criteria for performance evaluation and selection [12, 16, 37].

L. Marder (✉)
Universidade de Santa Cruz do Sul PPGSPI/UNISC, Santa Cruz do Sul–RS, Brazil
e-mail: lucmarder@yahoo.com.br

V. Pérez Herranz
Depto Ingeniería Química y Nuclear, Universidad Politécnica de Valencia, Valencia, Spain
e-mail: vperez@iqn.upv.es

A. Moura Bernardes et al. (eds.), *Electrodialysis and Water Reuse*,
Topics in Mining, Metallurgy and Materials Engineering,
DOI: 10.1007/978-3-642-40249-4_4, © Springer-Verlag Berlin Heidelberg 2014

This evaluation can be performed based on the transport number of counter ion through the membrane, t_j^m, which, according to Faraday's law is described by the Eq. (4.1):

$$t_j^m = \frac{z_j F J_j^m}{i} \tag{4.1}$$

where z_j is the charge of the counter-ion, F the Faraday constant, i the applied current density (mA·cm^{-2}) and J_j the counter-ion flow (mol·cm^{-2}·s^{-1}), which is defined by the Eq. (4.2):

$$J_j^m = \frac{V\left(C_j^f - C_j^0\right)}{A \cdot t'} \tag{4.2}$$

where V is the volume of the solution (L), C_j^f is the concentration of counter-ions at time t' (mol·L^{-1}), C_j^o is the initial concentration of counter-ions (mol·L^{-1}), A is the apparent area of the membrane (cm^2), t' is the time of the experiment (s).

Based on the counter-ion transport number, the membrane permselectivity can be obtained according to the relationship in Eq. (4.3):

$$P = \frac{t_j^m - t_j^s}{1 - t_j^s} \tag{4.3}$$

where $\mathrm{t_j^s}$ is the transport number of the counter-ion in solution [23].

The transport number of a given ion through the membrane can be obtained by measuring the change in concentration of this ion between the dilution compartment and the concentrating compartment as a function of time and applied current, as indicated by the Eqs. (4.1) and (4.2) [16, 37].

The membrane transport number in monovalent electrolytes can be obtained by the electromotive force method. In this method the transport number of the ion through the membrane is obtained from the potential difference measured between two reference electrodes arranged in two equivalent planes near the surfaces of the membrane, φ^m, according to Eq. (4.4) [37]:

$$\varphi^m = \frac{RT}{F}\left(2t_j^m - 1\right)\ln\frac{C_1}{C_2} \tag{4.4}$$

where R is the universal gas constant, F the Faraday constant, T the absolute temperature, C_1 and C_2 the concentrations of the electrolyte on both sides of the membrane. Usually diluted electrolyte solutions at different concentrations, such as $C_1 = 0.001$ and $C_2 = 0.005$ mol·L^{-1}, are employed.

Alternatively, the transport number of ions through the membranes can also be determined through the electrochemical method of chronopotenciometry [21], which is discussed in detail in Sect. 4.4.

4.2 Concentration Polarization and Limiting Current Density

In an electrodialysis process it is desirable to operate with the highest possible current densities in order to achieve the maximum ion flow per unit of membrane area. However, the operation levels are constrained by the concentration polarization [14]. This polarization occurs due to the difference in transport numbers of the ions in the membrane and the solution and can be described assuming diffusion boundary layers near the membrane surface.

While within the solution cations and anions carry approximately equal amounts of current, in the membrane, the current is almost exclusively carried by counter-ions due to the exclusion of co-ions. This behavior leads to a higher ion transport in the membrane than in the solution. As a result, concentration gradients are formed in the regions adjacent to the membrane, called diffusion boundary layers [14, 23], as can be seen in Fig. 4.1.

In the dilution compartment, the concentration decreases in relation to the bulk of the solution while in the concentration compartment, the concentration increases. Depending on the applied current density, the concentration at the membrane surface on the dilution side, can reach values close to zero.

The current density value where the concentration of the ions on the membrane surface on the dilution side approaches zero, is called the limiting current density, i_{lim}, and is defined by [14, 23]:

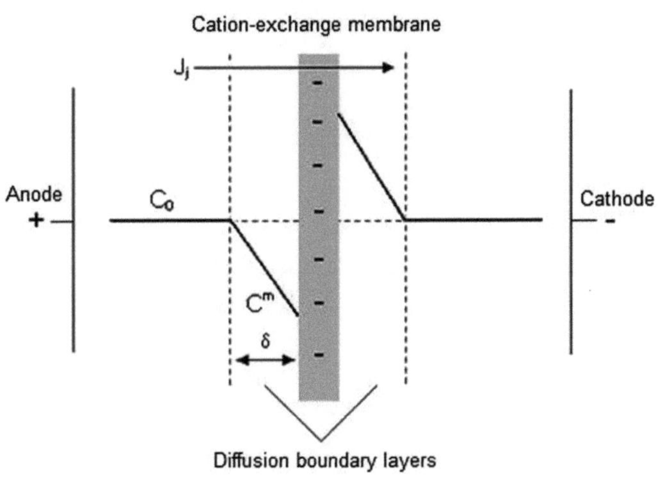

Fig. 4.1 Schematic diagram illustrating the polarization concentration: concentration gradients and diffusion boundary layers in a cation exchange membrane (J_j is the counter-ion flow across the membrane, C_0 is the concentration of salt within the solution, C^m is the salt concentration at the membrane surface, δ is the boundary layer thickness)

$$i_{\mathrm{lim}} = \frac{C_0 D z_j F}{\delta\left(t_j^m - t_j^s\right)} \tag{4.5}$$

where C_0 is the concentration of the salt within the solution, D is the salt diffusion coefficient in solution, z_j the charge of the counter-ion, F the Faraday constant, δ the thickness of the diffusion boundary layer, t_j^m and t_j^s the transport numbers of the counter-ion in the membrane and in the solution, respectively.

If in an electrodialysis process the limiting current density is exceeded, the efficiency of the process should drastically decrease due to an increase of the electrical resistance of the solution. In addition, such effects as the dissociation of water into H^+ and OH^- can occur, causing a change in the pH of the solutions on the surface of the membranes, causing additional operating problems, such as "scaling" (metal hydroxide deposits on the surface of the membranes). If the membranes are not resistant enough to acidic or alkaline environments, they could, therefore, also start to degrade [23].

The limiting current density can be determined by the method that relates the applied current density i, to the membrane potential obtained, φ^m [1, 3, 36]. When an ion selective membrane is placed between two electrolyte solutions, the response of the current density as a function of the membrane potential is similar to the curve presented in Fig. 4.2. Three characteristic regions can be observed on this curve: a quasi-ohmic variation of the current–voltage curves in the lower voltage range (region 1), followed by a plateau, from which the value of the limiting current is derived (region 2), and then an increase of the current density (region 3), which can be attributed to distinct phenomena accompanying the

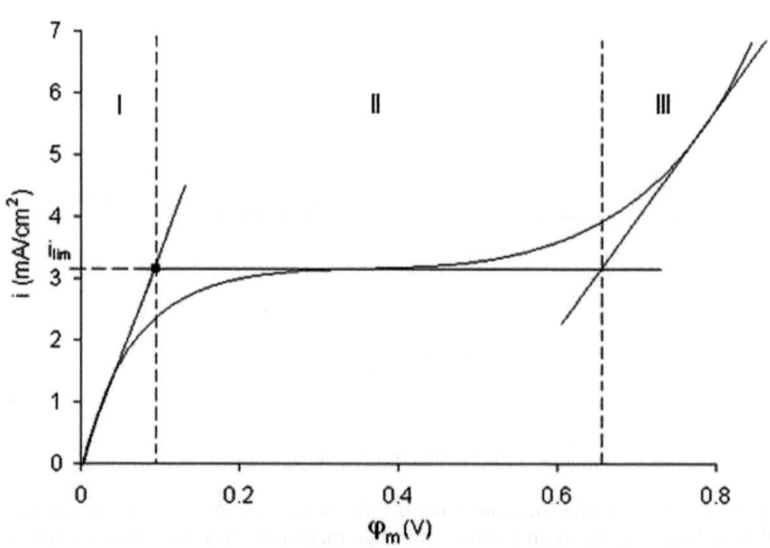

Fig. 4.2 Typical current-potential curve for an ion selective membrane

concentration polarization: (1) water dissociation, (2) exaltation of ions, (3) gravitational convection and (4) electro-convection [18, 34].

4.3 Phenomena Accompanying Concentration Polarization

According to the classical theory of concentration polarization, currents greater than the limiting current should not occur, since the concentration near the membrane approaches zero. However, in the case of ion exchange membranes, currents above the limiting current are observed and the third region is part of the characteristic current–potential curves [13]. Here we will discuss four phenomena that explain the emergence of these currents. The first two are related to the dissociation of water at the membrane/solution interface, while the other two are related to the mechanisms of convection.

4.3.1 Water Dissociation and Exaltation Effects

For several years it was believed that the current exceeding the limiting current was related to the additional current transport by H^+ and OH^-, formed by the dissociation of water. In addition, some studies have also shown that the dissociation of water may lead not only to the unproductive growth of current related to the transport by H^+ and OH^-, but also to some stimulation of the flow of ions within the solution to the membrane surface [18]. This phenomenon, known as the exaltation effect, is caused by the electric field created by the products of dissociation of water, H^+ and OH^-. This field interacts with the ionic species in the solution, such as OH^- ions with positive ionic species from the solution, causing an attraction effect, increasing the flow of such ionic species from the bulk of the solution to the surface of the membrane [18, 44, 45].

However, there is little evidence of water dissociation in cation exchange membranes, except when they are in contact with dilute solutions of $CrCl_2$ [33], $MgCl_2$ [9], $CoCl_2$ and $NiCl_2$ [35]. And the fact that this decoupling only occurs in anionic membranes containing ionizable functional groups $\equiv N$, $= NH$ and $-NH_2$ [9, 45], has generated a lot of discussion on the subject and new theories have emerged to explain this third region of the current–potential curve. This region cannot be attributed only to the water dissociation mechanism. Among these new theories, we can highlight those mechanisms related to convection: gravitational convection [44, 45] and electro-convection [18–20, 27–29, 45]. These theories are related to a disturbance of the diffusion boundary layer and, consequently, an additional flow of ionic species to the membrane surface.

4.3.2 Gravitational Convection

Gravitational convection occurs as the force of gravity acts on a fluid in which a density gradient is present. In the case of ion selective membranes, concentration gradients are formed near the membrane when the concentration polarization occurs and the concentration of ions in the diffusion boundary layer is lower compared to the concentration within the solution. This corresponds to a high strength and can lead to the production of Joule heating and the formation of temperature gradients near the membrane. Because of the concentration and temperature gradients, a density gradient should be present near the surface of the membrane when the concentration polarization occurs. The emergence of gravitational convection should cause a partial destruction of the diffusion boundary layer, thereby increasing the flow of ionic species towards the membrane surface. Heating the solution also leads to a certain increase in current, caused by an increase in the diffusion coefficient of the ions.

According to Zobolotsky et al. [44], this mechanism is significant when solutions are used that are not very dilute, for example, when 0.034 mol L^{-1} NaCl, are arranged between cationic and anionic membranes, separated by a relatively long distance and with a linear flow rate low. In this case, significant density gradients are formed between the diffusion boundary layer and the bulk solution, and gravitational convection is important. These authors also suggest that, under these conditions, the position of the membrane relative to Earth's gravitational field also exerts a significant influence on the development of this mechanism. When the diffusion boundary layer, where the depletion of ions occurs, is situated above a membrane that is positioned horizontally, the gravitational convection reaches its maximum. It is also important when the membrane is positioned vertically, however, it is absent when the diffusion boundary layer, where the depletion of ions occurs, is located below a membrane positioned horizontally. In the case of more dilute solutions with NaCl concentrations $\leq$0.002 mol L^{-1}, density gradients formed between the diffusion boundary layer and the bulk solution are negligible, and the gravitational convection should therefore also be negligible. For those cases where gravitational convection does not apply, the mechanism that must be contributing to the upper limit currents, according to Mishchuk [18], Mishchuk et al. [20], Zabolotsky et al. [45], and Rubinstein et al. [27–29], is electro-convection.

4.3.3 Electro-Convection

According to Mishchuk [18], when an excessively high electric field is applied, a region arises inside the diffusion boundary layer, called the space-charge region, which is characterized by a deviation from electroneutrality, i.e. the concentration of counter-ions and co-ions do not coincide. These space charges, distributed in a non-uniform way, move under the effect of the electric field, creating a pair of vortices capable of causing a solution mixture in the concentration polarization region. This may be interpreted in a similar way as gravitational convection: a partial destruction of the diffusion boundary layer resulting in an increased flow of ionic species from thein solution to the membrane surface [18, 28, 45].

4.4 Chronopotentiometry

Chronopotentiometry is an electrochemical characterization method that allows the monitoring of the change in membrane potential versus time for a given applied current density. The potential-time data obtained can be employed to obtain the values of the limiting current density and the transport number of ions through the ion exchange membranes. In addition, they provide important information about the heterogeneity of the membranes, the side effects caused by concentration polarization (water dissociation, gravitational convection and electro-convection), membrane fouling or scaling and the properties of membranes with multiple layers, particularly bipolar membranes [32, 43].

4.4.1 Principle

When a current density is applied between two electrodes arranged on the ends of an electrodialysis cell, comprised of an ion exchange membrane separating two compartments containing electrolyte solutions, and the membrane potential is recorded as a function of time, basically two types of chronopotentiometric curves can be obtained, depending on the applied current density [25, 32].

When current densities below the limiting current density ($i < i_{lim}$) are applied, curves similar to the one shown in Fig. 4.3 are obtained, where the ordinate shows the potential difference between two electrodes arranged near the membrane surface. Part of the curve corresponds to the ohmic drop of the system, including the membrane and the solution, and partly to concentration gradients [32].

When sufficiently high current densities, current densities that exceed the limit ($i > i_{lim}$), are applied, curves similar to the one shown in Fig. 4.4 are obtained [25].

The initial portion of this curve is characterized by three different sections: the first one (a), limited by point 1, is virtually vertical. Its height is equal to the ohmic drop of the system including the membrane and the solution. The second section (b) corresponds to a slight increase in potential until inflection point 2. This is due to the decrease in concentration of the dilute solution near the membrane, governed mainly by the processes of electro-diffusion. After passing the inflection point (section c) other mechanisms of mass transfer to the surface of the membrane, mainly convection, become important. Finally, the system reaches a steady state (point 3) where the potential does not vary anymore with time (section d). Sometimes, the portion of

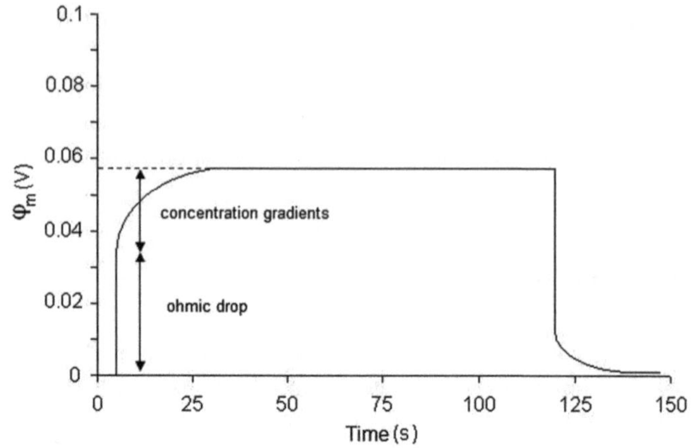

Fig. 4.3 Characteristic chronopotentiometric curve for current densities below the limiting current density

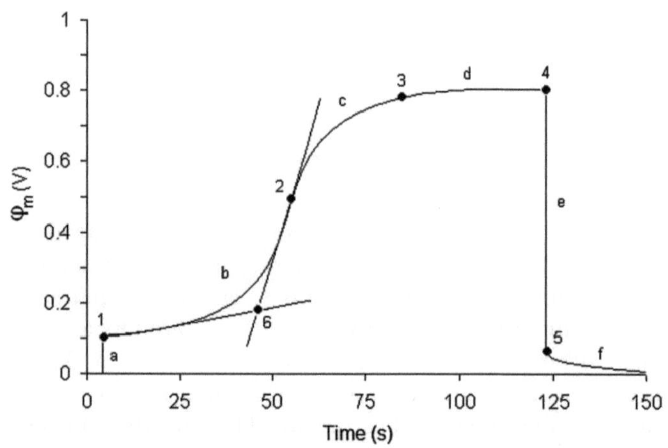

Fig. 4.4 Sections and feature points of a chronopotentiometric curve at current densities exceeding the limiting one [25]

these curves that corresponds to the processes that occur after the current interruption is also studied. The potential difference between the points 4 and 5 (section e) is equal to the potential drop over the polarized membrane system. The last section (f) describes the diffusion relaxation of the system [25].

An important characteristic presented by these curves, obtained at current densities that exceed the limiting current, is the transition time, τ, which may be determined as the point of intersection of two tangents formed by the regions of slow and fast potential change in section b (Fig. 4.4, point 6) [22, 25, 26, 30, 31, 38, 39], or as the inflection point (point 2) [4, 5, 6, 8, 10, 11, 14]. Theoretically, assuming a matter control by electro-diffusion and the absence of any convection, i.e. unlimited growth of the diffusion boundary layer near the membrane (semi-infinite diffusion process), the transition time, τ, can be calculated from the Sand equation [14, 25]:

$$\tau = \frac{\pi D}{4}\left(\frac{C_0 z_j F}{t_j^m - t_j^s}\right)^2 \frac{1}{i^2} \tag{4.6}$$

where D is the diffusion coefficient of the salt in solution, C_O is the concentration of the salt in solution, z_j is the charge of the counter ion, F the Faraday constant, i the current density, and t_j^m and t_j^s the transport number of the counter ion in the membrane and the solution, respectively.

4.4.2 Applications

The potential-time data obtained by chronopotentiometry can be employed to obtain different information about the transport of ions through ion exchange membranes, among which the following can be highlighted.

4.4.2.1 Limiting Current Density

The limiting current density can be determined by building the current-potential curve presented above (Fig. 4.2) and relating the applied current densities with corresponding values where the potential does not vary anymore with time [32, 34].

4.4.2.2 Transport Number of Ions Through the Membrane

The transport number of ions through the membranes can be obtained from the transition time obtained at current densities that exceed the limiting one, using the Sand equation (Eq. 4.5). According to this equation, the relationship between the transition time and the inverse of the squared current density ($1/i^2$), for different current densities that exceed the limiting current, should result in a straight line passing the origin. The value of the transport number of the ions through the membrane can be obtained from the slope of this straight line, since the values of the diffusion coefficient of the salt and the transport number of the ion in the solution are known. The diffusion coefficient data can be obtained from the equivalent conductivity data at infinite dilution using the Nernst–Einstein equation. The transport number of ions in the solution can also be calculated by employing the equivalent conductivity data [5, 7, 40].

4.4.2.3 Heterogeneity of Membranes. Conductive and Non-conductive Regions

If the transport number in the membrane, t_j^m, is available, information regarding the conducting area of the ion exchange membrane, ε, can be obtained by using chronopotentiometric data in the modified Sand's equation, suggested by Choi et al. [5]:

$$\tau = \frac{\varepsilon^2 \pi D}{4}\left(\frac{C_0 z_j F}{t_j^m - t_j^s}\right)^2 \frac{1}{i^2} \quad (4.7)$$

In a study developed with two structurally reinforced homogeneous cation exchange membranes (NEOSEPTA Selemion CMX and CMV) and a heterogeneous membrane (HJC produced by Jungsoo Hanguk Corporation), in contact with solutions of 0.02 mol L^{-1} NH_4Cl, the ε values found by Choi et al. [5], using the modified Sand's equation, were 93, 95 and 75 % for CMX, CMV and HJC, respectively. According to these authors, the values of ε are in agreement with the structural characteristics and mode of preparation of each membrane and, as a consequence, this equation is now used in a variety of studies. Among them, we can mention the work of Kang et al. [10] in which they determine the conducting fraction of ion exchange membranes based on poly-arylene-ether-sulfone (S-PES) according to the degree of sulfonation, and of commercial cation exchange membranes NEOSEPTA CM-1, CMX and CMB, in contact with NaCl solutions of 0.025 mol L^{-1}. The work of Nagarale et al. [22] can also be cited. In this work, the authors determined the conductive membrane fraction of heterogeneous cation exchange membranes based on polycarbonate and polysulphone as a function of the quantity of the resin added to the ion exchange membranes, in contact with solution of 0.01 mol L^{-1} NaCl.

4.4.2.4 Fouling by Organic Molecules and Mineral Deposits

Studies by Park et al. [24] and Shahi et al. [30] demonstrated that the observed behavior of the heterogeneous membranes can be employed to evaluate membrane fouling and scaling. The work done by Park et al. [24] shows that anionic ion selective membranes in contact with solutions of 0.01 mol L^{-1} KCl + 1 % bovine serum albumin (BSA) are susceptible to membrane fouling by deposition of BSA on the membrane surface. The obtained chonopotentiometric data shows this behavior in terms of a reduction in the fraction of the conductive region of the membrane, calculated by the modified sand equation, and also by a decrease in the limiting current density. Shahi et al. [30] evaluated the behavior of a cation exchange membrane modified by depositing copper or silver on the surface of the membrane. The chronopotentiometric experiments conducted with solutions of 0.01 mol L^{-1} NaCl, suggest that there is a decrease in the transition times obtained in accordance with the amount of metal deposited on the membrane surface.

With respect to the metal deposition on the surface of the membrane, Taki et al. [33], in a work carried out with solutions of 0.1 mol L^{-1} $CrCl_2$ + 0.1 mol L^{-1} HCl in contact with a Selemion CMV cation exchange membrane, observed an increase of the potential in the chronopotentiometric curves after the period of stabilization for high current densities, which coincides with the appearance of a precipitate of $Cr(OH)_3$ on the membrane surface. This increased potential can be interpreted as an increase in the membrane resistance due to the formed precipitate. The cause of precipitation can be related to the occurrence of the dissociation of water into H^+ ions and OH^-, which leads to an increase in pH at the membrane surface and, consequently, to the precipitation of the metal hydroxide.

4.4.2.5 Side Effects Associated with Concentration Polarization

Other studies have correlated the shape of the chronopotentiometric curves, obtained at excessively high current densities, to the phenomena associated to concentration polarization, such as water dissociation, gravitational convection and electro-convection, which were discussed previously.

Pismenskaya et al. [25], in studies carried out with Selemion AMX, MA-40 and MA-41anion exchange membranes in contact with NaCl solutions of 0.01 and 0.1 mol L^{-1}, and Belova et al. [2] in studies with Selemion AMX, MA-40 and MA-40-13 membranes in contact with solutions of NaCl 0.005 and 0.1 mol L^{-1}, observed that, depending on the characteristics of the membranes and experimental conditions, the chronopotentiomteric curves may have a maximum potential before reaching the steady state or even oscillations present in the region where the potential reaches its equilibrium state. These authors related the first behavior to a reduced region of concentration polarization caused by gravitational convection or a decrease in the electrical resistance of the membrane and the solution due to the appearance of H^+ and OH^- ions produced by water dissociation or by local heating in the region of concentration polarization (temperature increase). Regarding the second behavior, Pismenskaya et al. [25] and Belova et al. [2] relate these oscillations to hydrodynamic instabilities caused by the mechanism of electro-convection. Krol et al. [15], in a work studying NEOSEPTA CMX membranes in contact with NaCl solutions of 0.1 mol L^{-1}, also observed and related these instabilities to the mass transfer mechanism. In a study by Belova et al. [2], chronopotentiometric curves, where membrane potential continues to grow slowly in the region where the equilibrium state should already have been reached, are presented. These curves are obtained for MA-40 anionic membranes, characterized by the presence of the functional groups =NH and ≡N. They attribute the observed behavior to a gradual blockage of functional groups by OH^- ions generated by the dissociation of water. According to these authors, the electrical conductivity of this membrane decreases dramatically when it starts to form OH^-, which explains the potential growth.

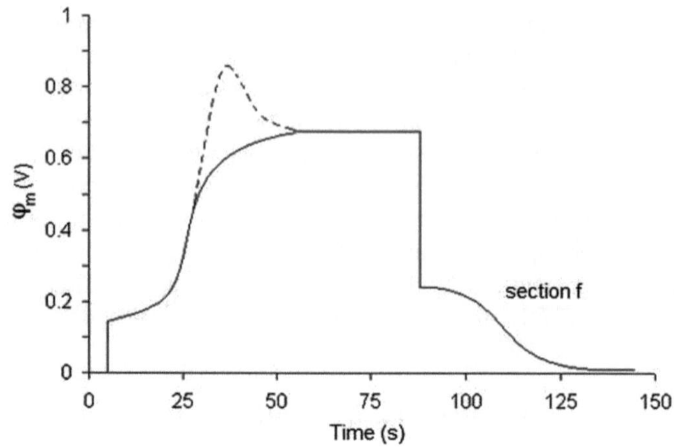

Fig. 4.5 Typical chronopotentiometric curve of a bipolar membrane obtained at current densities exceeding the limiting one. The maximum potential (*dashed curve*) is observed only at high current densities [41]

4.4.2.6 Transport Properties for Bipolar Membranes

Wilhelm et al. [41, 42] have used this technique to study the transport phenomena in bipolar membranes. In this case, chronopotentiomteric curves, like those shown in Fig. 4.5, are obtained for current densities exceeding the limiting one. One can see that the initial region of the curve is similar to the curves obtained for monopolar membranes, but the region corresponding to the diffusion relaxation system (section f), is considerably different. These authors have attributed this behavior to the recombination of $H^+$and OH^-, formed by water dissociation at the junction of the bipolar membrane.

Marder et al. [17], studying the transport properties of manganese, cobalt, copper, nickel and zinc through the IONICS 67-HMR-412 monopolar cation exchange membrane, observed a similar behavior in bipolar membranes when nickel and cobalt solutions were electrodialysed at high current densities. These authors observed that under these conditions a great layer of precipitate is formed on the surface of the membrane and suggest that at currents exceeding the limit, a water splitting phenomena occurs when cobalt and nickel ions are employed and a bipolar layer is formed in the membrane/solution interface.

References

1. Barragán VM, Bauzá CR (2002) Current-voltage curves for a cation-exchange membrane in methanol—water electrolyte solutions. J Colloid Interface Sci 247:138–148. doi:10.1006/jcis.2001.8065

2. Belova EI, Lopatkova GY, Pismenskaya ND et al (2006) Effect of anion exchange membrane surface properties on mechanisms of overlimiting mass transfer. J Phys Chem B 110:13458–13469. doi:10.1021/jp062433f
3. Chamoulaud G, Bélanger D (2005) Modification of ion exchange membrane used for separation of protons and metallic cations and characterization of the membrane by current-voltage curves. J Colloid Interface Sci 281:179–187. doi:10.1016/j.jcis.2004.08.081
4. Choi E-Y, Strathmann H, Park JM et al (2006) Characterization of non-uniformly charged ion exchange membranes prepared by plasma-induced graft polymerization. J Membr Sci 268:165–174. doi:10.1016/j.memsci.2005.06.052
5. Choi J-H, Kim S-H, Moon S-H (2001) Heterogeneity of ion—exchange membranes: the effects of membranes heterogeneity on transport properties. J Colloid Interface Sci 241:120–126. doi:10.1006/jcis.2001.7710
6. Choi Y-J, Kang M-S, Kim S-H et al (2003) Characterization of LPDE/polystyrene cation exchange membranes prepared by monomer sorption and UV radiation polymerization. J Membr Sci 223:201–215. doi:10.1016/S0376-7388(03)00339-9
7. Lide DR (1998) Handbook of chemistry and physics. CRC press, New York
8. Ibanez R, Stamatialis DF, Wessling M (2004) Role of membrane surface in concentration polarization at cation—exchange membranes. J Membr Sci 239:119–128. doi:10.1016/j.memsci.2003.12.032
9. Jialin L, Yazhen Y, Changying Y et al (1998) Membrane catalytic deprotonation effects. J Membr Sci 147:247–256. doi:10.1016/S0376-7388(98)00126-4
10. Kang M-S, Choi Y-J, Choi I-J et al (2003) Electrochemical characterization of sulfonated poly (arylene ether sulfone) (S-PES) cation exchange membranes. J Membr Sci 216:39–53. doi:10.1016/S0376-7388(03)00045-0
11. Kang M-S, Choi Y-J, Lee H-J et al (2004) Effects of inorganic substances on water splitting in ion exchange membranes. Electrochemical characteristics of ion exchange membranes coated with iron hydroxide/oxide and silica sol. J Colloid Interface Sci 273:523–532. doi:10.1016/j.jcis.2004.01.050
12. Koter S (2001) Transport number of counterions in ion—exchange membranes. Sep Purf Technol 22–23:643–654. doi:10.1016/S1383-5866(00)00160-X
13. Krol JJ (1997) Monopolar and bipolar membranes: mass transporte limitations. PhD thesis, University of Twente, Print Partners Ipskamp, Enschede, The Netherlands, p 173
14. Krol JJ, Wessling M, Strathmann H (1999) Chronopotentiometry and overlimiting ion transport through monopolar ion exchange membranes. J Membr Sci 162:155–164. doi:10.1016/S0376-7388(99)00134-9
15. Krol JJ, Wessling M, Strathmann H (1999) Concentration polarization with monopolar ion—exchange membranes: current—voltage curves and water dissociation. J Membr Sci 162:145–154. doi:10.1016/S0376-7388(99)00133-7
16. Lteif R, Dammak L, Larchet C et al (2001) Détermination du nombre de transport d'un contre-ion dans une membrane échangeuse d'ions en utilisant la méthode de la pile de concentration. Eur Polym J 37:627–639. doi:10.1016/S0014-3057(00)00163-4
17. Marder L, Ortega-Navarro EM, Pérez-Herranz V et al (2006) Evaluation of transition metals transport properties through a cation exchange membrane by chronopotentiometry. J Membr Sci 284(1–2):267–275. doi:10.1016/j.memsci.2006.07.039
18. Mishchuk NA (1998) Electro—osmosis of the second kind near the heterogeneous ion exchange membrane. Colloids Surf A 140(1–3):75–89. doi:10.1016/S0927-7757(98)00216-7
19. Mishchuk NA (1998) Perspectives of the electrodialysis intensification. Desalination 117:283–296. doi:10.1016/S0011-9164(98)00120-9
20. Mishchuk NA, Koopal LK, Gonzalez-Caballero F (2001) Iintensification of electrodialysis by applying a non—stationary electric field. Colloids Surf A 176(2–3):195–212. doi:10.1016/S0927-7757(00)00568-9
21. Nagarale RK, Gohil GS, Shahi VK (2006) Recent developments on ion—exchange membranes and electromembranes processes. Adv Colloid Interface Sci 119(2–3):97–130. doi:10.1016/j.cis.2005.09.005

22. Nagarale R, Shahi VK, Thampy SK et al (2004) Studies on electrochemical characterization of polycarbonate and polysulfone bases heterogeneous cation—exchange membranes. React Funct Polym 61:131–138
23. Noble RD, Stern SA (1995) Membrane separations technology—principles and applications. Elsevier, Amsterdan
24. Park J-S, Choi J-H, Yeon K-H et al (2006) An approach to fouling characterization of an ion exchange membranes using current—voltage relation and electrical impedance spectroscopy. J Colloid Interface Sci 294(1):129–138. doi:10.1016/j.jcis.2005.07.016
25. Pismenskaya N, Sistat P, Huguet P et al (2004) Chronopotenciometry applied to the study of ion transfer through anion exchange membranes. J Membr Sci 228(1):65–76. doi:10.1016/j.memsci.2003.09.012
26. Ramachandraiah G, Ray P (1997) Electroassisted transport phenomenon of strong and weak electrolytes across ion exchange membranes: chronopotentiometric study on deactivation of anion exchange membranes by higher homologous monocarboxylates. J Phys Chem B 101:7892–7900. doi:10.1021/jp9701698
27. Rubinstein I, Staude E, Kedem O (1998) Role of the membrane surface in concentration polarization at ion—exchange membrane. Desalination 69:101–114. doi:10.1016/0011-9164(88)80013-4
28. Rubinstein I, Zaltzman B, Kederm O (1997) Electric fields in and around ion—exchange membranes. J Membr Sci 125(1):17–21. doi:10.1016/S0376-7388(96)00194-9
29. Rubinstein I, Zaltzman B, Pretz J et al (2002) Experimental verification of the electroosmotic mechanism of overlimiting conductance through a cation—exchange electrodialysis membrane. Rus J Electrochem 38:853–863
30. Shahi VK, Prakash R, Ramachandraiah G et al (1999) Solution—membrane equilibrium at metal—deposited cation exchange membranes: chronopotentiometry characterization of metal-modified membranes. J Colloid Interface Sci 216(1):179–184
31. Shahi VK, Thampy SK, Rangarajan R (2002) Chronopotentiometric studies on dialytic properties of glycine across ion—exchange membranes. J Membr Sci 203(1–2):43–51. doi:10.1016/S0376-7388(01)00745-1
32. Sistat P, Pourcelly G (1997) Chronopotentiometric response of an ion—exchange membrane in the underlimiting currente—range. Transport phenomena within the diffusion layers. J Membr Sci 123:121–131. doi:10.1016/S0376-7388(96)00210-4
33. Taky M, Pourcelly G, Gavach C et al (1996) Chronopotentiometric response of a cation—exchange membrane in contact with chromium (III) solutions. Desalination 105(3):219–228. doi:10.1016/0011-9164(96)00079-3
34. Taky M, Pourcelly G, Lebon F et al (1992) Polarization phenomena at the interfaces between an electrolyte solution and an ion—exchange membrane. Part 1: ion transfer with a cation exchange membrane. J Electroanal Chem 336(1–2):171–194. doi: 10.1016/0022-0728(92)80270-E
35. Tanaka Y (2002) Water dissociation in ion—exchange membrane electrodialysis. J Membr Sci 203(1–2):227–244. doi:10.1016/S0376-7388(02)00011-X
36. Tanaka Y (2003) Concentration polarization in ion—exchange membrane electrodialysis—the events arising in a flowing solution in a desalting cell. J Membr Sci 216(1–2):149–164. doi:10.1016/S0376-7388(03)00067-X
37. Tourreuil V, Rossignol N, Bulvestre G et al (1998) Détermination de la séléctivite d'une membrane échangeuse d'ion: confrontation entre le flux de diffusion et le nombre de transport. Eur Polym J 34(10):1415–1421. doi:10.1016/S0014-3057(97)00288-7
38. Volodina E, Pismenskaya N, Nikonenko V et al (2005) Ion transfer across ion—exchange membranes with homogeneous and heterogeneous surfaces. J Colloid Interface Sci 285(1):247–258. doi:10.1016/j.jcis.2004.11.017
39. Vyas PV, Ray P, Adhikary SK et al (2003) Studies of the effect of variation of blend ratio on permselectivity and heterogeneity of ion exchange membranes. J Colloid Interface Sci 257(1):127–134. doi:10.1016/S0021-9797(02)00025-5
40. Moore WJ (1972) Physical chemistry. Prentice Hall, New Jersey

41. Wilhelm FG, Van der Vegt NFA, Strathmann H et al (2002) Comparison of bipolar membranes by means of chronopotentiometry. J Membr Sci 199(1–2):177–190. doi:10.1016/S0376-7388(01)00696-2
42. Wilhelm FG, Van der Vegt NFA, Wessling M et al (2001) Chronopotentiometry for the advanced current–voltage characterisation of bipolar membranes. J Electroanalyt Chem 502(1–6):152–166. doi:10.1016/S0022-0728(01)00348-5
43. Xu T (2005) Ion—exchange membranes: state of their development and perspective. J Membr Sci 263(1–2):1–29. doi:10.1016/j.memsci.2005.05.002
44. Zabolotsky VI, Nikonenko VV, Pismenskaya ND (1996) On the role of gravitational convection in the transfer enhancement of slat ions in the course of dilute solution electrodialysis. J Membr Sci 119:171–181. doi: dx.doi.org/10.1016/0376-7388(96)00121-4
45. Zabolotsky VI, Nikonenko VV, Pismenskaya ND et al (1998) Coupled transport phenomena in overlimiting current electrodialysis. Sep Purif Technol 14(1–3):255–267. doi:10.1016/S1383-5866(98)00080-X

Chapter 5
Ionic Membranes

Carlos A. Ferreira, Franciélli Müller and Franco D. R. Amado

Abstract Ionic membranes made of both natural and synthetic polymers have been developed for over 50 years and currently comprise a wide range of applications. In comparison with traditional industrial separation processes, such as adsorption, extraction, and distillation, membrane technology has inherent advantages (e.g., less reagent and solvent consumption, no chemical additives, and recyclability), leading to reduced energy consumption while the performance of the separation processes increases. The performance of ion exchange membranes can be assessed based on the following aspects: flux and selectivity, cost of production, thermal and chemical stability, and mechanical strength. These properties are essentially related to the chemical and physical nature of polymers. In the last few decades, the design of materials that are able to combine high ionic conductivity and durability, good mechanical strength, reduced permeability, and low cost for high-volume production has become one the major challenges in the field of new polymer materials for electrodialysis applications. Therefore, the main objective of the present chapter is to give a brief summary of the different materials, properties and characterization of ionic membranes as well as of the synthesis of electrodialysis membranes.

C. A. Ferreira (✉) · F. Müller
Programa de Pós Graduação em Engenharia de Minas, Metalúrgica e de Materiais (PPGE3M), Universidade Federal do Rio Grande do Sul (UFRGS), Porto Alegre–RS, Brazil
e-mail: ferreira.carlos@ufrgs.br

F. Müller
e-mail: franciellim@yahoo.com.br

F. D. R. Amado
Departamento de Ciências Exatas e Tecnológicas, Universidade Estadual de Santa Cruz (UESC), Ilhéus–BA, Brazil
e-mail: franco.amado@gmail.com

A. Moura Bernardes et al. (eds.), *Electrodialysis and Water Reuse*, Topics in Mining, Metallurgy and Materials Engineering,
DOI: 10.1007/978-3-642-40249-4_5,

5.1 Membrane Materials

One of the advantages of the use of conductive polymers in ion selective membranes is the fact that transport properties can be reversibly altered. The most important research on the subject was realized by Murray and Burgmayer [15], who developed the idea of "ion-gate membranes". They showed that an oxidized polypyrrole film presented permeability for simple ions, different from the reduced polypyrrole films.

5.1.1 Polymer Intrinsically Conducting (PICs) Membranes

Conducting, polymeric based membranes have been produced and studied in laboratories. Membranes with conductive polymers can be obtained by the in situ polymerization technique, which consists of a chemical synthesis of the conductive polymer on membranes previously impregnated with a monomer solution (or oxidant). It can also be found membranes prepared by mechanical mixing of conductive and conventional polymers.

5.1.1.1 Membranes Obtained by In Situ Polymerization

Sata et al. [47, 49–51] have obtained modified cationic or anionic membranes through in situ polymerization. Pyrrole or aniline was polymerized over commercial NEOSEPTA membranes, assembled by Tokuyama Soda Co. These membranes have a PVC support and poly(styrene-co-divinylbenzene) based ion-change resins. The used cationic membranes were of strong acid type (SO_3H^- groups) and the anionic membranes were of strong alkali type (trimethylbenzylammonium groups).

Nagarale et al. [42] also have chemically polymerized aniline in one of sides of commercial membranes. The membrane's performance was evaluated using electrodialysis with simple solutions or by a mixture of electrolytes to observe the electromigration between solutions. They used solutions like Na_2SO_4, $CaCl_2$ and $CuCl_2$, taking NaCl as reference. The results showed that the permselectivity for cations of the membranes with Polyaniline (PAni) was $Na^+ > Ca^{2+} > Cu^{2+}$ [42].

Other authors have also used the technique of in situ polymerization to modify commercial membranes. Among them Ferreira et al. [54], who used Selemion® CMT membranes made of sulfonated poly(styrene-co-divinylbenzene) with a PVC screen as support and polymerized pyrrole with $FeCl_3$ as an oxidant. They also have polymerized aniline with ammonium persulfate as the oxidant. The results for ion exchange capacity and percentage extraction for Cl^- anions and SO_4^{2-} were similar to those obtained for the commercial membranes.

Sazou et al. coated a surface of stainless steel with a commercial Nafion 117 membrane with deposited polyaniline in order to inhibit pitting corrosion in a chloride-containing environment. The anti-corrosive strategy is based on modifying the stainless steel interface with cation selective films of Nafion and PAni, avoiding the contact of chloride ions with the steel surface. The authors have shown promising results with respect to corrosion inhibition [53].

5.1.1.2 Membrane Obtained Through PICs and Conventional Polymer Mixture

The use of polyaniline is hampered by its infusibility and impossibility of solubilize the polymer in the conducting state in organic solvents commonly used in conventional polymer processing.

Great progress was made in this field by Heeger et al. [16], with the development of a PAni solubilizing method. This method is based on "functionalized protonic acids" (FPA), which form a complex with the polymer and, at the same time, promote doping and solubility for the polyaniline in several organic solvents. The proton's acid reacts with the imine nitrogen of polyaniline and converts from the emeraldine base form to the emeraldine salt-conductive form. The protoned polyaniline functional groups are compatible with non-polar and weakly polar solvents. The dodecylbenzenesulfonic acid (DBSA) and camphorsulfonic acid (CSA) are the most used.

Djurado et al. have shown that these dopants make easier the PAni processing by the induced plasticization by doping, be known as "plastdopants". Plastdoped PAni is protoned PAni with high electrical conductivity and excellent mechanical properties [21, 55].

The conductive blend assembling containing polymers intrinsically conductive, including polyaniline, through mechanical mixture with conventional polymers, has been widely studied [8, 20]. These blends are promising, because they provide the polymeric material with controlled conductivity. Furthermore, they are economically viable, present excellent assembling and a mechanical resistance comparable to conventional polymers [20, 38].

A variety of PAni-made blends has been obtained, including blends made with polyethylene, polystyrene, poly(methyl methacrylate), polycarbonates, polypropylene, polyurethane and high impact polystyrene [38].

One of the methods for preparing polyaniline and conventional polymer blends is mechanical mixing using a counter rotating mixer or a twin screw extruder [40]. In such cases, the result will depend strongly on the conductivity and rheological properties of miscibility of the blend components [20].

The mixing temperature and screw speed are variables that must be adjusted according to the polymer to be mixed. The temperature should be high enough to permit flow of the material, and low enough to avoid any degradation of the polymers.

Another frequently used method is to dissolve the blend components in a solvent, followed by its evaporation. In this method, phase segregation may occur during evaporation of the solvent, since the solubility of the polymers may be different.

Davies and co-workers [19], following the model of Heeger [16] have solubilized doped polyaniline with an excess of dodecylbenzenesulfonic acid (DBSA), which in addition to doping the polymer also plasticizes, making polyaniline more soluble in non-polar solvents such as toluene. Subsequently, they dispersed solubilized PAni in SBS or poly(methyl methacrylate) (PMMA), getting blends with good dimensional stability and conductivity greater than 10 S/cm.

Several researchers have used the polyaniline dissolution in a compatible solvent with conventional polymers to obtain conductive blends for several uses [9, 31, 43, 62].

5.1.1.3 Membrane Obtained by Interpenetrating Polymer Nets

Interpenetrating polymer nets (IPNs) can be defined as a combination of two polymers with a net structure, or which at least one has been synthesized or reticulated in the presence of the other polymer [56]. Figure 5.1 shows the difference between the physical structures of IPNs schematically.

Polyurethane is one of the most commonly used polymers for the formation of an IPN. It is formed from the polycondensation reaction of polyols with isocyanates. Polyester polyols and castor oil are among the polyols used in polyurethane synthesis (Fig. 5.2).

Castor oil has been widely used as polyol in polyurethane synthesis due to its stability under different conditions of pressure and temperature. It is also an alternative to petrochemically derived polyols [18].

Some authors have devoted themselves to the study of IPNs [17, 44, 60, 61], including Siddaramaiah and colleagues [34], who have developed composite polyurethane IPNs from castor oil and polyaniline, aiming to improve the

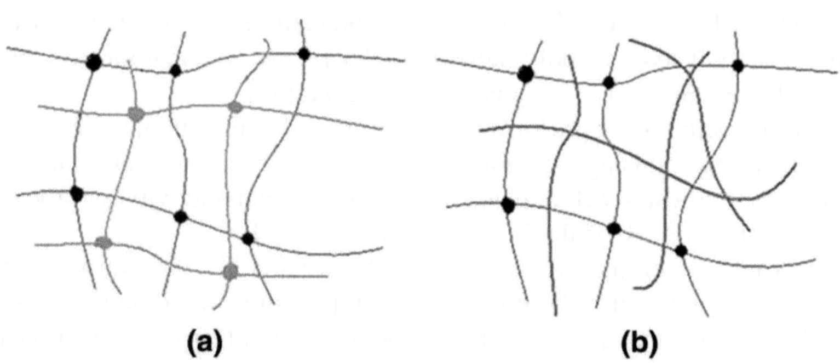

Fig. 5.1 Idealized structures of an IPN (**a**) and a semi-IPN (**b**)

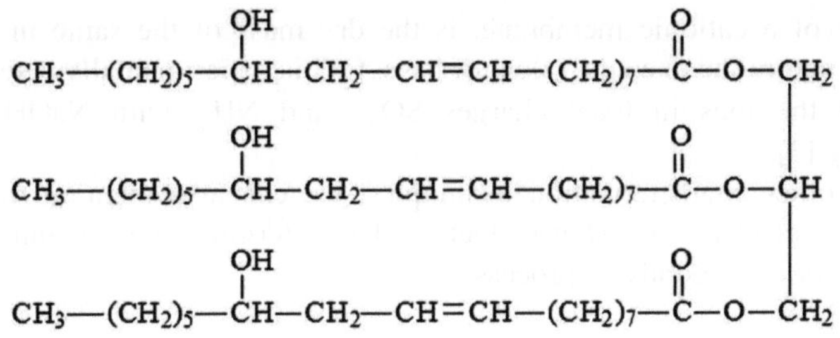

Fig. 5.2 Chemical structure of castor oil

processability and properties of cross-linked polymers that contain conductive polymers, since their solubility is limited. The IPNs were produced by the mixture of castor oil and diisocyanate, to which polyaniline was added later.

5.2 Membrane Characterization

Ion selective membranes used in electrodialysis can be classified in terms of their mechanical and electrical properties, their selectivity and their chemical stability. The electric charge of a membrane can be determined qualitatively by the use of an indicator. This test consists of applying a droplet of a solution with 0.05 % methylene blue and metilorange in a sample. If it presents a yellow color, the membrane is anionic. If it presents a blue color, the membrane is cationic [50].

The mechanical characterization of a membrane includes determining its thickness, swelling, dimensional stability, tension resistance and hydraulic permeability [51].

The swelling capacity of a membrane determines not only its dimensional stability, but affects its selectivity, electrical resistance and hydraulic permeability. The "swelling" of a membrane depends on the nature of the polymeric material and the cross-link density. Usually it is expressed by the difference in mass of the dry and wet membrane.

The chemical stability of a membrane is its ability to resist decay in contact with various corrosive environments. This is evaluated based on ASTM D 543-95. The membranes are evaluated after 24, 48 and 168 h, analyzing changes in color, texture, luster, decomposition, cracks, holes, blisters, bending and tack. This rating is estimated by visual comparison with new membranes [3].

Previous studies have shown that membranes obtained from HIPS and PAni had not been deteriorated and that no significant change occurred in its structure in acid and alkaline media. In the same strongly oxidizing medium, the untreated membrane is degraded and cannot be used again [54].

Another frequently used test to characterize membranes is the ion exchange capacity (IEC) test, which is determined by the equivalent mass of the membrane.

In the case of a cationic membrane, is the dry mass of the same in acid form required to neutralize one equivalent of base. CTI is experimentally determined by titration of the ions in fixed charges SO_3^- and NH_4^+ with NaOH and HCl respectively [3].

Among other characterization techniques, we can also mention the permselectivity and electrical resistance, factors that determine the amount of energy required in an electrodialysis process.

5.2.1 Heterogeneous Ion Exchange Membranes

Heterogeneous membranes can be subdivided into reinforced and non-reinforced membranes. The first uses conventional polymers as support, such as polyvinylchloride (PVC). They can be produced in various ways:

- By dispersing the ion exchange resin or polymer that presents exchange groups in powder form in a polymer solution, with subsequent evaporation of the solvent and creation of the membrane;
- By mixing a base polymer with an ion exchange resin in an extruder, for example, and subsequent pressing to form the membrane.

The affinity of the resin for various cations is of significant influence on the efficiency of the process because it is harder to move an ion that has a high affinity for the resin than an ion that has low affinity [59]. Those ions that have a higher affinity for the membrane will have greater difficulty passing through.

The cationic resin affinity for cations is:

$$Li^+ < Na^+ < K^+ < Mg^{2+} < Ca^{2+} < Al^{3+} < Fe^{3+}$$

The anionic resin affinity for anions is:

$$HSiO_3^- \ll F^- < HCO_3^- < Cl^- < SO_4^{2-}$$

5.2.2 Homogeneous Ion Exchange Membrane

Homogeneous membranes for electrodialysis are assembled from the functional monomer or by functionalization of a polymeric film by sulfonation. These membranes can be reinforced or auto supported.

Ion selective membranes may be made homogeneous by sulfochlorination and amination of polyethylene films and these membranes have low electrical resistance combined with high mechanical strength and excellent permselectivity [57].

5.2.3 Special Membranes

Due to the widespread development of electrodialysis techniques on an industrial scale, membranes with different properties have been developed. Among these membranes, we can find:

Membranes that are permselective to monovalent ions: These are membranes that separate monovalent ions of a solution containing different valence ion mixture. Their main application is the separation of chloride sodium of salt water.

Anti-fouling anion selective membranes: To avoid the fouling problem, some organizations adjust the degree of polymer reticulation. Ion selective membranes based on aliphatic polymers show a reduction in the obstruction by organic ions when compared with membranes based on aromatic polymers.

Membranes that are selective to fluoride-carbon cations: Most conventional ion selective membranes are degraded by oxidant agents, especially at high temperatures. Nafion® membranes developed by DuPont present excellent chemical and thermal stability for use in chloride-soda plants.

Bipolar membranes: Recently, bipolar membranes have earned attention as an efficient tool for the production of acids and alkalis from their salts by water dissociation in the interface between the anionic and cationic layers under an electric current. Bipolar membranes, in general, exhibit unsatisfactory chemical stability at high pH and a high electrical resistance [57].

5.3 Membrane Properties

In the field of separation science and technology, membrane processes are currently being studied in numerous applications with practical interests. The electrodialysis system is determined by the properties of the membranes used in the various applications [27].

Generally, membrane properties include the equilibrium between physical–mechanical properties (exchange capacity, water content, sorption properties, thickness, thermal and chemical stability); transport properties (electroconductivity, diffusion and electroosmotic permeability, transport numbers of ions) and structural characteristics measured by various physical methods (X-ray, spectral and optical methods that enable the investigation of the structure of membranes at different scales) [2, 13, 26, 29, 32, 35, 45].

The ion exchange membrane properties used in electrodialysis depend on the polymeric matrix and the concentration of fixed charges. Membrane properties for these wide applications includes the comprehensive studies, where the most desired properties required for the successful ion exchange membranes are mechanical and electrical properties, their permselectivity and their chemical stability [11, 24, 30, 33]. These properties are essentially related to the chemical and physical nature of polymers. In the last few decades, the design of materials has permitted to combine high ionic conductivity and durability, good mechanical strength with reduced permeability, and low costs for high-volume production has become one of the major challenges in the field of new polymer materials for electrodialysis applications [12, 36, 54].

5.3.1 Permselectivity of Membranes

Permselectivity between ions with the same charge is generally evaluated by the equivalence of specific ions permeated through the membrane when one equivalent of a standard cation or anion permeates [29]. An ion exchange membrane should be highly permeable to counter-ions, but should be impermeable to co-ions [33]. When ions with the same charge have to be separated, separation by molecular sieving based on the polymer network is firstly thought. That is, it is thought that smaller ions permeate selectively through a membrane with a highly crosslinked structure when compared to larger ions [48].

5.3.2 Mechanical Properties of Membranes

Membranes should be mechanically strong and should have a low degree of swelling or shrinkage in transition from dilute to concentrated ionic solutions [27].

The mechanical properties of ion selective membranes include their thickness, swelling, dimensional stability, tensile strength and hydraulic permeability. The tensile strength test permits the measurement of the Young's modulus E, which gives information about the rigidity of the polymer membrane and its tensile strength R_m, which is the limit stress value supported by the membrane [23].

The swelling capacity of the membrane determines not only its dimensional stability, but also affects its selectivity, electrical resistance and hydraulic permeability. The swelling of the membrane depends on the nature of the polymeric material, the ion exchange capacity and the crosslink density.

Hydraulic permeability measurements provide information on the transport of components through the membrane and on the hydrostatic pressure. The presence of micro holes in an ion exchange membrane not only increases the hydraulic permeability, but invalidates their application.

5.3.3 Chemical Stability of Membranes

Ion exchange membranes should have high chemical resistance, and ideally should be stable to various chemical products over the entire pH-range and in the presence of oxidizing agents.

5.3.4 Electrical Resistance of Membranes

The electrical resistance of ion exchange membranes is one of the factors that determine the energy requirements of electrodialysis systems [57] and it is an important property that is expressed as electrical resistance per unit area (Ωcm^2). Although electrical resistance is usually measured in sodium chloride or potassium chloride solutions [29, 58], this membrane property changes remarkably with the species of counter-ions. Usually, a standardized solution and standard conditions are used to compare the electrical resistance of various membranes.

5.3.5 Ion Exchange Capacities of Membranes

The method employed to determine the ion exchange capacity depends on the kinds of ion exchange groups that are used by an ion exchange membrane. However, most cation exchange membranes have sulfonic acid or/and carboxylic acid groups, while anion exchange membranes have tertiary amino groups or quaternary ammonium groups. To determine the dissociation of the ion exchange groups with the pH of the solution, a pH titration curve of the membrane can be determined.

The ion exchange capacity of an ion exchange membrane is an important parameter because the ionic transport properties of the membranes depend on the amount and species of the ion exchange groups.

To determine the ion exchange capacity of cation exchange membranes, the membranes are equilibrated in a 1 M HCl solution for 72 h. After that, they are removed from the solution, and the excess of acid is eliminated by washing the films repeatedly with distilled water. Next, membranes are immersed in 1 M NaCl to exchange protons by Na^+. The amount of H^+ in these three solutions is determined by titration with 0.005 M NaOH to a phenolphthalein end point. IEC is expressed as milliequivalent per gram of membrane (meq/g-dry) [3, 22] (Eq. 5.1).

$$IEC = \frac{V_{NaOH} . M_{NaOH}}{W_{dry}} \tag{5.1}$$

where V_{NaOH} and M_{NaOH} are the blank-corrected volume (mL) and molar concentration (mol/L) of NaOH solution, respectively.

To determine the ion exchange capacity to anion exchange membranes, a membrane sample is immersed for 48 h in 1 M KOH solution after it has been dipped in pure water for 24 h, to convert the membrane into the OH^- form. The membrane is then washed several times with distilled water, then equilibrated in this medium to remove traces of the KOH solution. The membrane is then immersed for 24 h in a 10^{-2} M HCl solution. The anion exchange capacity is determined from the decrease in acidity, which is obtained through back titration after neutralization of the membrane. The ion exchange capacity is determined with the following relation (Eq. 5.2):

$$IEC = \frac{\left(\eta_{HCl,0} - \eta_{HCl,af.neut.}\right)}{m} \quad (5.2)$$

where $\eta_{HCl,0}$ is the HCl quantity in the initial solution, $\eta_{HCl,af.neut.}$ the HCl quantity in the solution after neutralization of the membrane and m the mass of the dry membrane [1].

The properties of some commercially available ion exchange membranes are listed in Table 5.1 [57, 63].

The parameters determining the membrane properties listed above often have the opposing effect. For instance, a high degree of crosslinking improves the mechanical strength, but increases the electrical resistance. More ionic charges in the membrane matrix lead to a low electrical resistance. In general, however, they cause a high degree of swelling combined with poor mechanical stability. It is often difficult to optimize the properties of ion exchange membranes because the parameters determining the different properties often act contrary to each other [14, 33]. The properties of ion exchange membranes are determined by two parameters, namely the basic material they are made from and the type and concentration of the fixed ionic moiety. The basic material determines to a large extent the mechanical, chemical, and thermal stability of the membrane.

5.4 Electrodialysis Membranes

Today, advances have been made in the synthesis and modification of polymeric ion exchange membranes. Nevertheless, extensive studies are still under way to develop new materials and methods for membrane modification and to produce more complex membrane structures with a wide range of functionalities and desired operating performance.

As far as their chemical structure is concerned, ion exchange membranes are very similar to normal ion exchange resins. From a chemical point of view, these resins would make excellent membranes of high selectivity and low resistivity. The difference between membranes and resins arises largely from the mechanical requirements of the membrane process. Unfortunately, ion exchange resins are mechanically weak, cation resins tend to be brittle, and anion resins are soft. They are dimensionally unstable due to the varying amounts of water imbibed into the gel under different conditions. Changes in electrolyte concentration, in the ionic form, or in temperature may cause major changes in the water uptake and hence in the volume of the resin. These changes can be tolerated in small spherical beads. But, in large sheets that have been cut to fit an apparatus, they are unacceptable. Thus, it is generally not possible to use sheets of material that have been prepared in the same way as a bead of resin. However, the most common solution to this problem is the preparation of a membrane that is backed by a stable reinforcing material that gives the necessary strength and dimensional stability [33].

Table 5.1 Properties of some commercial ion exchange membranes [57, 63]

Membrane	Structure properties	Thickness (mm)	IEC (meq/g)	Area resistance (Ωcm^2)
Asahi Chemical Industry Co. Japan				
Aciplex K-192	CEM	0.13–0.17	–	1.6–1.9
Aciplex-501SB	CEM	0.16–0.20	–	1.5–3.0
Aciplex A-192	AEM	>0.15	–	1.8–2.1
Aciplex-501SB	AEM	0.14–0.18	–	2.0–3.0
Aciplex A201	AEM	0.22–0.24	–	3.6–4.2
Aciplex A221	AEM	0.17–0.19	–	1.4–1.7
Asahi Glass Co. Ltd., Japan				
Selemion CMV	CEM	0.13–0.15	–	2.0–3.5
Selemion AMV	AEM	0.11–0.15	–	1.5–3.0
Selemion ASV	AEM	0.11–0.15	–	2.3–3.5
Selemion DSV	AEM	0.13–0.17	–	–
DuPont Co., USA				
Nafion NF-112	CEM	0.051	–	–
Nafion NF-1135	CEM	0.089	–	–
Nafion NF-115	CEM	0.127	–	–
Nafion N-117	CEM	0.183	0.90	1.5
FuMA-Tech GmbH, Germany				
FKS	CEM	0.090–0.110	0.90	2–4
FKB	CEM	0.100–0.115	0.80	5–10
FK-40	CEM	0.035–0.045	1.20	1.00
FKD	CEM	0.040–0.060	1.00	1.00
FAS	AEM	0.100–0.120	1.10	2–4
FAB	AEM	0.090–0.110	0.80	2–4
FAN	AEM	0.090–0.110	0.80	2–4
FAA	AEM	0.080–0.100	1.10	2.4

(continued)

Table 5.1 (continued)

Membrane	Structure properties	Thickness (mm)	IEC (meq/g)	Area resistance (Ωcm^2)
FAD	AEM	0.08	1.30	1.2
Ionics Inc., USA				
CR61-CMP	CEM	0.58–0.70	2.2–2.5	11
CR67-HMR	CEM	0.53–0.65	2.1–2.45	7.0–11.0
CR67-HMP	–	–	–	–
AR103QDP	AEM	0.56–0.69	1.95–2.20	14.5
AR204SZRA	AEM	0.48–0.66	2.3–2.7	6.2–9.3
AR112-B	AEM	0.48–0.66	1.3–1.8	20–28
MEGA a.s., Czech Republic				
Ralex MH-PES	AEM	0.55 (Dry)	1.80	<8
Ralex AMH-5E	AEM	0.7 (Dry)	1.80	<13
Ralex CM-PES	CEM	0.45 (Dry)	2.20	<9
Ralex CMH-5E	CEM	0.6 (Dry)	2.20	<12
PCA Polymerchemie Altmeier GmbH, Germany				
PC 100 D	AEM	0.08–0.1	1.2 Quat.	5
PC 200 D	AEM	0.08–0.1	1.3 Quat.	2
PC Acid 35	AEM	0.08–0.1	1.0 Quat.	–
PC Acid 70	AEM	0.08–0.1	1.1 Quat.	–
PC Acid 100	AEM	0.08–0.1	0.57 Quat.	–
PC-SK	CEM	–	–	–
PC-SA	AEM	–	–	–
Solvay S.A., Belgium				
Morgane CDS	CEM	0.130–0.170	1.7–2.2	0.7–2.1
Morgane CRA	CEM	0.130–0.170	1.4–1.8	1.8–3.0
Morgane ADP	AEM	0.130–0.170	1.3–1.7	1.8–2.9
Morgane AW	AEM	0.130–0.170	1.0–2.0	0.9–2.5

(continued)

Table 5.1 (continued)

Membrane	Structure properties	Thickness (mm)	IEC (meq/g)	Area resistance (Ωcm^2)
Tokuyama Co., Japan				
Neosepta CM-1	CEM	0.13–0.16	2.0–2.5	0.8–2.0
Neosepta CM-2	CEM	0.12–0.16	1.6–2.2	2.0–4.5
Neosepta CMX	CEM	0.14–0.20	1.5–1.8	2.0–3.5
Neosepta CMS	CEM	0.14–0.17	2.0–2.5	1.5–3.5
Neosepta CMB	CEM	0.22–0.26	–	3.0–5.0
Neosepta AM-1	AEM	0.12–0.16	1.8–2.2	1.3–2.0
Neosepta AM-3	AEM	0.11–0.16	1.3–2.0	2.8–5.0
Neosepta AMX	AEM	0.12–0.18	1.4–1.7	2.0–3.5
Neosepta AHA	AEM	0.18–0.24	–	3.0–5.0
Neosepta ACM	AEM	0.10–0.13	1.4–1.7	3.5–5.5
Neosepta ACS	AEM	0.12–0.20	1.4–2.0	3.0–6.0
Neosepta AFN	AEM	0.15–0.18	2.0–3.5	0.2–1.0
Neosepta AFX	AEM	0.14–0.17	1.5–2.0	0.7–1.5
Tianwei Membrane Co. Ltd., China				
TWEDG	AEM	0.16–0.21	1.6–1.9	3–5
TWDDG	AEM	0.18–0.23	1.9–2.1	<3
TWAPB	AEM	0.16–0.21	1.4–1.6	5–8
TWANS	AEM	0.17–0.20	1.2–1.4	6–10
TWAHP	AEM	0.20–0.21	1.2–1.4	<2
TWAEDI	AEM	0.18–0.21	1.6–1.8	6–8
TWCED	CEM	0.16–0.18	1.4–1.6	2–4
TWCDD	CEM	0.16–0.18	1.6–2.0	2–4
TWCEDI	CEM	0.16–0.18	1.2–1.4	5–8

The measurement conditions to determine the area resistance varied with companies: Asahi Chemical Co., 0.5 M NaCl at 25 °C; Asahi Glass Co. Ltd., 1 kHz AC in a 0.5 M NaCl solution; DuPont Co., 0.5 M NaCl at 25 °C; FuMA-Tech, GmbH, 0.6 M NaCl at 25 °C; PCA Polymerchemie Altmeier, GmbH, 1 N KCl; Solvay S.A., 1 kHz AC in 10 g/l NaCl at 25 °C, except AW which is measured in a 1 M H_2SO_4, HCl or HNO_3 solution at 25 °C; Tokuyama Co., 0.5 N NaCl at 25 °C; Tianwei Co., 0.1 N NaCl at 17 °C

Ion exchange membranes are traditionally classified into anion exchange membranes and cation exchange membranes, depending on the type of ionic groups attached to the membrane matrix. Cation exchange membranes are generally films of polystyrene-co-divinylbenzene, polysulfone, fluoro-carbonated or other conjugated polymers that have been functionalized to allow ions to pass through their structure under an applied potential. Cation exchange membranes have a very acidic content of negatively charged groups, such as $–SO_3^-$, $–COO^-$, $–PO_3^{2-}$, $–PO_3H^-$, $–C_6H_4O^-$, etc., that are fixed to the membrane backbone and allow the passage of cations, but reject anions. Similarly, anion exchange membranes have a very basic content of positively charged groups, such as $-NH_3^+$, $-NRH_2^+$, $-NR_2H^+$, $-NR_3^+$, etc., that are fixed to the membrane backbone and allow the passage of anions, but reject cations [25, 28, 57, 63]. These strong basic anion exchange membranes present quaternary ammonium exchange groups and weak basic anion exchange membranes with groups derived from secondary or primary amines [46, 54].

These different ionic groups have significant effects on the selectivity and electrical resistance of the ion exchange membrane. The sulfonic acid group, e.g. –SO_3^- is completely dissociated over nearly the entire pH-range, while the carboxylic acid group $–COO^-$ is virtually undissociated in the pH-range <3. Similarly the quaternary ammonium group $-NR_3^+$ is completely dissociated over the entire pH-range, while the secondary ammonium group $-NR_2H^+$ is only weakly dissociated. Accordingly, ion exchange membranes are referred to as being weakly or strongly acidic or mostly basic in character. Commercially available cation exchange membranes have $–SO_3^-$ or $–COO^-$ groups, and anion exchange membranes contain mainly $-NR_3^+$ groups.

Ion exchange membranes can also be divided into two major categories according to their structure and preparation procedure: homogeneous membranes and heterogeneous membranes.

5.4.1 Preparation of Homogeneous Ion Exchange Membranes

To synthesize homogeneous ion exchange membranes, various approaches are available to introduce ionic groups, which are classified in three categories:

- A monomer containing a moiety that either is or can be made anionic or cationic exchange groups is copolymerized with a non-functionalized monomer to eventually form an ion exchange membrane.
- A polymer film is modified by introducing ionic characters either directly, by the grafting of a functional monomer, or indirectly, by the grafting of a nonfunctional monomer followed by a functionalization reaction.
- An anionic or cationic moieties are introduced on a polymer or polymer blends followed by the dissolving of the polymer and its casting on a film.

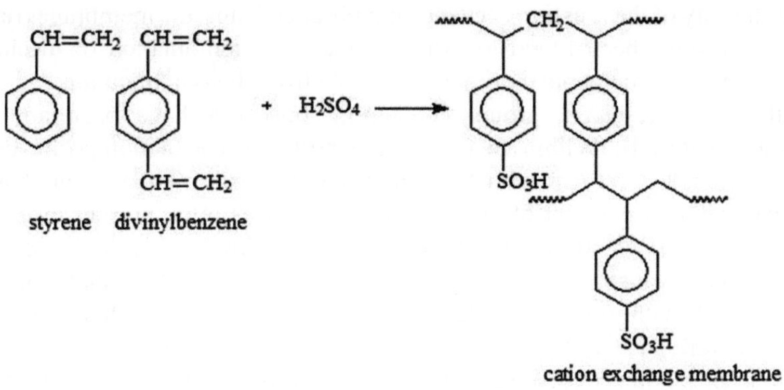

Fig. 5.3 Chemical structure of a cation exchange membrane

If the membrane is prepared from a monomer, then styrene and divinylbenzene are most commonly utilized as the neutral starting material for a traditional hydrocarbon type ion exchange membrane for industrial use. A basic anion exchange membrane is usually prepared in two steps, chloromethylation and quaternary amination. A cation exchange membrane is prepared by sulfonation.

There are several references on the preparation of such membranes [10, 37, 39, 52]. These are the best-known membranes in the literature for the application of electrodialysis. The cation exchange membrane (Fig. 5.3) is obtained according to the illustrated reaction scheme [33, 63].

In a first step styrene is partially polymerized with divinylbenzene using benzoyl peroxide as an initiator for the polymerization. The polymer is partially cross-linked and obtained as a block, which is cut into slices. In a second step it is sulfonated with concentrated sulfuric acid at room temperature. The obtained membranes show high ion exchange capacity and low electrical resistance, but their mechanical strength is insufficient to be used without a proper support material.

However, by adding linear polymers, such as polyvinylchloride, polyethylene-styrene-butadiene rubber, etc. to a mixture of styrene, divinylbenzene, and a plasticizer, a pasty mixture is obtained. The mixture is then coated on a backing fabric, such as a net, and covered on both sides with a glass plate or a separating plastic sheet of poly(vinyl alcohol) or polyethylene terephthalate. By heating the composite layer, vinyl monomers are copolymerized. The obtained film is then sulfonated with concentrated sulfuric acid as described earlier. The result is a reinforced ion exchange membrane with excellent electrical and mechanical properties.

Müller [41] developed blends of polystyrene (PS) and high impact polystyrene (HIPS) with poly(styrene-ethylene-butylene) (SEBS) triblock sulfonated copolymer, to fabricate cation exchange membranes by solvent-casting and subsequent thermal treatment to obtain homogeneous packing densities for electrodialysis applications. Furthermore, a conducting polymer, polyaniline (PAni) doped with camphorsulfonic acid (CSA), was added to enhance both the electrical conductivity and the stability of these cation exchange membranes. Thus, the membranes reported in this work have been found to exhibit a valuable combination of mechanical, chemical and electrochemical properties for electrodialysis applications.

In the same research, group, conducting polymers have been applied as membrane materials. Within this field, the performance of composite cationic membranes made of PAni with high-impact polysterene (HIPS/PAni) has been examined for the treatment of zinc solutions, which is often present in industrial effluents, with the electrodialysis process [3, 5–7]. Amado [4] developed cation exchange membranes produced by mixing polyurethane (PU) and polyaniline (PAni) doped with p-toluenesulfonic acid (pTSA), and camphorsulfonic acid (CSA). The influence of the polyaniline concentration and the dopant nature on the membrane properties were investigated. The results for zinc extraction obtained using these membranes were compared to those obtained using the commercial Nafion 450 membrane.

A homogeneous anion exchange membrane (Fig. 5.4) can be obtained by introducing a positively charged quaternary amine group to a preformed polymer by a chloromethylation procedure followed by amination with a tertiary amine according to the following reaction scheme [33]:

Membranes with good electrical and mechanical properties were obtained by additional polymerization with divinylbenzene or by first polymerizing the monochloromethylstyrene with divinylbenzene and then treating the resulting polymeric film with trimethylamine. The membrane structures described above, and their preparation, are just two examples. There are many variations of the

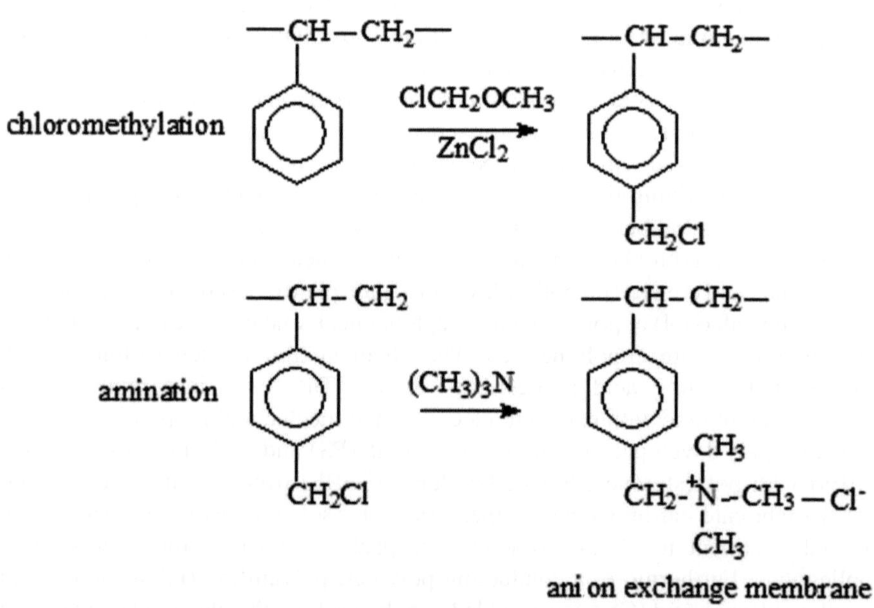

Fig. 5.4 Chemical structure of an anion exchange membrane

basic preparation procedure, which result in slightly different products. Instead of styrene, styrene substitutes, such as methylstyrene or phenylacetate, are often used and instead of divinylbenzene, monomers such as divinylacetylene or butadiene are used. Furthermore, instead of sulfonic acid, phoposphoric or arsenic acid is introduced to the cross-linked polystyrene. However, most of these membranes have no, or only very little commercial relevance.

5.4.2 Preparation of Heterogeneous Ion Exchange Membranes

Ion exchange membranes with a heterogeneous structure consist of fine ion exchange resin particles embedded in an inert binder such as polyethylene, phenolic resins, or polyvinylchloride. Heterogeneous ion exchange membranes can easily be prepared by different methods [33, 57]:

- An ion exchange powder is mixed with a dry binder polymer and molded into a sheet under appropriate conditions of pressure and temperature.
- Ion exchange particles are mixed with a binder polymer, which is brought to a semifluid state by adding a plasticizer or by heating and then extruding it as a sheet.
- Ion exchange particles are dispersed in a solution containing a dissolved film-forming binder, then the mixture is cast into a film and the solvent evaporated.
- Similarly, ion exchange particles are dispersed in a partially polymerized binder polymer and cast into a film, then the polymerization is completed.

References

1. Agel E, Bouet J, Fauvarque JF (2001) Characterization and use of anionic membranes for alkaline fuel cells. J Power Sour 101(2):267–274. doi:10.1016/s0378-7753(01)00759-5
2. Alhadidi A, Kemperman AJB, Schippers JC et al (2011) The influence of membrane properties on the silt density index. J Membr Sci 384(1–2):205–218. doi:10.1016/j.memsci.2011.09.028
3. Amado FDR, Gondran E, Ferreira JZ et al (2004) Synthesis and characterisation of high impact polystyrene/polyaniline composite membranes for electrodialysis. J Membr Sci 234(1–2):139–145. doi:10.1016/j.memsci.2004.01.017
4. Amado FDR, Rodrigues LF Jr, Rodrigues MAS et al (2005) Development of polyurethane/polyaniline membranes for zinc recovery through electrodialysis. Desalination 186(1–3):199–206. doi:10.1016/j.desal.2005.05.019
5. Amado FDR, Rodrigues LF, Forte MMC et al (2006) Properties evaluation of the membranes synthesized with castor oil polyurethane and polyaniline. Polym Eng Sci 46(10):1485–1489. doi:10.1002/pen.20602
6. Amado FDR, Rodrigues MAS, Morisso FDP (2008) High-impact polystyrene/polyaniline membranes for acid solution treatment by electrodialysis: preparation, evaluation, and chemical calculation. J Colloid Interface Sci 320(1):52–61. doi:10.1016/j.jcis.2007.11.054

7. Amado FDR, Rodrigues MAS, Bertuol DA et al (2009) The effect of production method on the properties of high impact polystyrene and polyaniline membranes. J Membr Sci 330(1–2):227–232. doi:10.1016/j.memsci.2008.12.065
8. Anand J, Palaniappan S, Sathyanarayana DN (1998) Conductive polyaniline blends and composites. Prog Polym Sci 23:993–1018
9. Barra GMO, Leyva ME, Soares BG, Sens M (2002) Solution-cast blends of polyaniline–DBSA with EVA copolymers. Synth Met 130(3):239–245. doi:10.1016/s0379-6779(02)00115-7
10. Bayramoglu G, Senkal FB, Celik G, Arica MY (2007) Preparation and characterization of sulfonyl-hydrazine attached poly(styrene-divinylbenzene) beads for separation of albumin. Colloids Surf A Physicochem Eng Asp 294(1–3):56–63. doi:10.1016/j.colsurfa.2006.07.043
11. Berezina NP, Kononenko NA, Dyomina OA et al (2008) Characterization of ion exchange membrane materials: properties vs structure. Adv Colloid Interface Sci 139(1–2):3–28. doi:10.1016/j.cis.2008.01.002
12. Bertran O, Curcó D, Torras J et al (2010) Field-induced transport in sulfonated poly(styrene-co-divinylbenzene). Membr Macromol 43(24):10521–10527. doi:10.1021/ma102500w
13. Bouzek K, Moravcová S, Schauer J et al (2010) Heterogeneous ion selective membranes: the influence of the inert matrix polymer on the membrane properties. J Appl Electrochem 40(5):1005–1018. doi:10.1007/s10800-009-9974-3
14. Burggraaf AJ (2004) Chapter 3 Preparation and characterization of ion-exchange membranes. Strathmann H (ed) Membrane Science and Technology, vol 9. Elsevier, PP 89–146. http://www.sciencedirect.com/science/article/pii/S0927519304800317. doi: 10.1016/S0927-5193(04)80034-2
15. Burgmayer P, Murray RW (1982) An ion gate membrane: electrochemical control of ion permeability through a membrane with an embedded electrode. J Am Chem Soc 104: 6139–6140
16. Cao Y, Smith P, Heeger AJ (1992) Counter-ion induced processibility of conducting polyaniline and of conducting polyblends of polyaniline in bulk polymers. Synth Met 48(1): 91–97. doi:10.1016/0379-6779(92)90053-l
17. Chwang C-P, Liu C-D, Huang S-W et al (2004) Synthesis and characterization of high dielectric constant polyaniline/polyurethane blends. Synth Met 142(1–3):275–281. doi:10.1016/j.synthmet.2003.09.012
18. Costa HMd, Ramos VD, Abrantes TAS et al (2004) Efeito do óleo de mamona em composições de borracha natural contendo sílica. Effect of castor oil in natural rubber compositions containing silica. Polímeros 14:46–50
19. Davies SJ, Ryan TG, Wilde C et al (1995) Processable forms of conductive polyaniline. Synth Met 69(1–3):209–210. doi:10.1016/0379-6779(94)02418-x
20. De Paoli M-A (1997) Conductive polymer blends and composites. In: Ltda JWS (ed) Handbook of organic conductive molecules in polymers, vol 2. Conductive polymers: synthesis and eletrical properties
21. Djurado D, Bée M, Gonzalez M et al (2003) Molecular dynamics in plastic conducting compounds of polyaniline. Chem Phys 292(2–3):355–361. doi:10.1016/s0301-0104(03)00121-6
22. Ferreira CA, Casanovas J, Rodrigues MAS et al (2010) Transport of metallic ions through polyaniline-containing composite membranes. J Chem Eng Data 55(11):4801–4807. doi:10.1021/je1004033
23. Ghalloussi R, Garcia-Vasquez W, Bellakhal N et al (2011) Ageing of ion exchange membranes used in electrodialysis: investigation of static parameters, electrolyte permeability and tensile strength. Sep Purif Technol 80(2):270–275. doi:10.1016/j.seppur.2011.05.005
24. Heitner-Wirguin C (1996) Recent advances in perfluorinated ionomer membranes: structure, properties and applications. J Membr Sci 120(1):1–33. doi:10.1016/0376-7388(96)00155-x

25. Higa M, Nishimura M, Kinoshita K et al (2012) Characterization of cation exchange membranes prepared from poly(vinyl alcohol) and poly(vinyl alcohol-b-styrene sulfonic acid). Intern J Hydrog Energy 37(7):6161–6168. doi:10.1016/j.ijhydene.2011.06.003
26. Hnát J, Paidar M, Schauer J et al (2012) Polymer anion selective membranes for electrolytic splitting of water. Part II: enhancement of ionic conductivity and performance under conditions of alkaline water electrolysis. J Appl Electrochem 42(8):545–554. doi:10.1007/s10800-012-0432-2
27. Hosseini SM, Madaeni SS, Khodabakhshi AR (2010) Preparation and characterization of ABS/HIPS heterogeneous cation exchange membranes with various blend ratios of polymer binder. J Membr Sci 351(1–2):178–188. doi:10.1016/j.memsci.2010.01.045
28. Hosseini SM, Madaeni SS, Heidari AR et al (2011) Preparation and characterization of polyvinyl chloride/styrene butadiene rubber blend heterogeneous cation exchange membrane modified by potassium perchlorate. Desalination 279(1–3):306–314. doi:10.1016/j.desal.2011.06.022
29. Hosseini SM, Gholami A, Madaeni SS et al (2012) Fabrication of (polyvinyl chloride/cellulose acetate) electrodialysis heterogeneous cation exchange membrane: characterization and performance in desalination process. Desalination. doi:10.1016/j.desal.2012.07.028
30. Hosseini SM, Madaeni SS, Heidari AR et al (2012b) Preparation and characterization of ion selective polyvinyl chloride based heterogeneous cation exchange membrane modified by magnetic iron–nickel oxide nanoparticles. Desalination 284:191–199. doi:10.1016/j.desal.2011.08.057
31. Jousseaume V, Morsli M, Bonnet A et al (1998) Electronic structure of conducting polyaniline blends. Opt Mater 9(1–4):480–483. doi:10.1016/s0925-3467(97)00088-8
32. Kanakasabai P, Deshpande AP, Varughese S (2012) Novel polymer electrolyte membranes based on semi-interpenetrating blends of poly(vinyl alcohol) and sulfonated poly(ether ether ketone). J Appl Polym Sci n/a-n/a. doi:10.1002/app.37749
33. Kariduraganavar MY, Nagarale RK, Kittur AA et al (2006) Ion exchange membranes: preparative methods for electrodialysis and fuel cell applications. Desalination 197(1–3):225–246. doi:10.1016/j.desal.2006.01.019
34. Kendaganna BK, Siddaramaiah S, Somashekarappa H et al (2004) Structure-properties relationship in polyaniline-filled castor oil based chain extended polyurethanes. Polym Eng Sci (44):772
35. Khongnakorn W, Wisniewski C, Pottier L et al (2007) Physical properties of activated sludge in a submerged membrane bioreactor and relation with membrane fouling. Sep Purif Technol 55 (1):125–131. doi:10.1016/j.seppur.2006.11.013
36. Li Y, Shimizu H (2006) Morphological investigations on the nanostructured poly(vinylidene fluoride)/polyamide 11 blends by high-shear processing. Eur Polym J 42(12):3202–3211. doi:10.1016/j.eurpolymj.2006.08.014
37. Li J, Ichizuri S, Asano S et al (2005) Proton exchange membranes prepared by grafting of styrene/divinylbenzene into crosslinked PTFE membranes. Nucl Instrum Methods Phys Res Sect B 236(1–4):333–337. doi:10.1016/j.nimb.2005.03.281
38. Lima JR, Schreiner C, Berton R et al (1998) Charge injection from polyaniline-poly methylmethacrylate blends into poly(p-phenylene vinylene). J Appl Phys 84:1445–1448
39. Mahmoud Nasef M, Saidi H (2003) Preparation of crosslinked cation exchange membranes by radiation grafting of styrene/divinylbenzene mixtures onto PFA films. J Membr Sci 216(1–2):27–38. doi:10.1016/s0376-7388(03)00027-9
40. Martins CR, De Paoli M-A (2005) Antistatic thermoplastic blend of polyaniline and polystyrene prepared in a double-screw extruder. Eur Polym J 41(12):2867–2873. doi:10.1016/j.eurpolymj.2005.06.016
41. Müller F, Ferreira CA, Franco L et al (2012) New Sulfonated Polystyrene and Styrene–Ethylene/Butylene–Styrene Block Copolymers for Applications in Electrodialysis. J Phys Chem B 116(38):11767–11779. doi:10.1021/jp3068415

42. Nagarale RK, Gohil GS, Shahi VK et al (2004) Preparation and electrochemical characterization of cation- and anion exchange/polyaniline composite membranes. J Colloid Interface Sci 277(1):162–171. doi:10.1016/j.jcis.2004.04.027
43. Namazi H, Kabiri R, Entezami A (2002) Determination of extremely low percolation threshold electroactivity of the blend polyvinyl chloride/polyaniline doped with camphorsulfonic acid by cyclic voltammetry method. Eur Polym J 38(4):771–777. doi:10.1016/s0014-3057(01)00232-4
44. Nayak P, Mishra DK, Parida D et al (1997) Polymers from renewable resources. IX. Interpenetrating polymer networks based on castor oil polyurethane poly(hydroxyethyl methacrylate): Synthesis, chemical, thermal, and mechanical properties. J Appl Polym Sci 63(5):671–679. doi:10.1002/(sici)1097-4628(19970131)63:5<671::aid-app15>3.0.co;2-x
45. Park S-J, Cheedrala R, Diallo M et al (2012) Nanofiltration membranes based on polyvinylidene fluoride nanofibrous scaffolds and crosslinked polyethyleneimine networks. J Nanopart Res 14(7):1–14. doi:10.1007/s11051-012-0884-7
46. Rodrigues MAS, Korzenovski C, Gondran E et al (2006) Evaluation of changes on ion selective membranes in contact with zinc-cyanide complexes. J Membr Sci 279(1–2):140–147. doi:10.1016/j.memsci.2005.11.045
47. Sata T (1993) Properties of composite membranes formed from ion exchange membranes and conducting polymers. 4. Change in membrane resistance during electrodialysis in the presence of surface-active agents. J Phys Chem 97(26):6920–6923. doi:10.1021/j100128a029
48. Sata T (1994) Studies on ion exchange membranes with permselectivity for specific ions in electrodialysis. J Membr Sci 93(2):117–135. doi:10.1016/0376-7388(94)80001-4
49. Sata T, Yamaguchi T, Matsusaki K (1996) Preparation and properties of composite membranes composed of anion exchange membranes and polypyrrole. J Phys Chem 100(41):16633–16640. doi:10.1021/jp961024o
50. Sata T, Ishii Y, Kawamura K et al (1999) Composite membranes prepared from cation exchange membranes and polyaniline and their transport properties in electrodialysis. J Electrochem Soc 146(2):585–591. doi:10.1149/1.1391648
51. Sata T, Sata T, Yang W (2002) Studies on cation exchange membranes having permselectivity between cations in electrodialysis. J Membr Sci 206(1–2):31–60. doi:10.1016/s0376-7388(01)00491-4
52. Savari S, Sachdeva S, Kumar A (2008) Electrolysis of sodium chloride using composite poly(styrene-co-divinylbenzene) cation exchange membranes. J Membr Sci 310(1–2):246–261. doi:10.1016/j.memsci.2007.10.049
53. Sazou D, Kosseoglou D (2006) Corrosion inhibition by Nafion®-polyaniline composite films deposited on stainless steel in a two-step process. Electrochim Acta 51(12):2503–2511. doi:10.1016/j.electacta.2005.07.033
54. Scherer R, Bernardes AM, Forte MMC et al (2001) Preparation and physical characterization of a sulfonated poly(styrene-co-divinylbenzene) and polypyrrole composite membrane. Mater Chem Phys 71(2):131–136. doi:10.1016/s0254-0584(00)00515-0
55. Sniechowski M, Djurado D, Dufour B et al (2004) Direct analysis of lamellar structure in polyaniline protonated with plasticizing dopants. Synth Met 143(2):163–169. doi:10.1016/j.synthmet.2003.11.010
56. Sperling LH (1981) Interpenetrating Polymer Networks and Related Materials. Plenum, New York
57. Strathmann H (1995) Chapter 6 electrodialysis and related processes. In: Richard DN, Stern SA (eds) Membrane Science and Technology, vol 2. Elsevier, pp 213–281. doi: 10.1016/S0927-5193(06)80008-2
58. Sun Koo J, Kwak N-S, Hwang TS (2012) Synthesis and properties of an anion exchange membrane based on vinylbenzyl chloride–styrene–ethyl methacrylate copolymers. J Membr Sci 423–424:239–301. doi:10.1016/j.memsci.2012.08.024
59. Topper EB, Wirth LF (1956) Ion exchange resins. In: Schubert FCNAJ (ed) Ion exchange technology. Academic Press Inc., New York, pp 7–26

60. Tran NB, Pham QT (1997) Castor oil-based polyurethanes: 2. Tridimensional polyaddition in bulk between castor oil and diisocyanates—gelation and determination of Fw(OH). Polym 38(13):3307–3314. doi:10.1016/s0032-3861(96)00890-7
61. Trân NB, Vialle J, Pham QT (1997) Castor oil-based polyurethanes: 1. Structural characterization of castor oil—nature of intact glycerides and distribution of hydroxyl groups. Polym 38(10):2467–2473. doi:10.1016/s0032-3861(96)00791-4
62. Wessling B (1998) Dispersion as the link between basic research and commercial applications of conductive polymers (polyaniline). Synth Met 93(2):143–154. doi:10.1016/s0379-6779(98)00017-4
63. Xu T (2005) Ion exchange membranes: State of their development and perspective. J Membr Sci 263(1–2):1–29. doi:10.1016/j.memsci. 2005.05.002

Chapter 6
Electrodialysis in Water Treatment

Andréa Moura Bernardes and Marco A. S. Rodrigues

Abstract This chapter focuses on the uses of electrodialysis and specially electrodialysis reversal for the treatment of brackish and groundwater to produce drinking water. Over the last 10–15 years, numerous advances in membrane and system technologies have made EDR an especially attractive technology, both in terms of performance and cost effectiveness. Many applications of EDR technology can be found around the world and different applications are shown here.

6.1 Introduction

Water, the universal solvent, can dissolve to some extent all organic and inorganic components found in the environment. The composition of water, therefore, will change through natural or anthropogenic impacts.

While the natural quality of drinking water depends primarily on the geology and soils of the catchment, other factors such as land use and disposal of pollutants are also important. In general, impermeable rocks, such as granite, are associated with turbid, soft, slightly acidic and naturally colored waters. Groundwater resources associated with hard rock geology are localized and small, so supplies come mainly from surface waters, such as rivers and impounding reservoirs. By contrast, permeable rocks such as chalk, limestone and sandstone produce clear, hard waters (high in Ca^{+2} and Mg^{+2}), which are slightly alkaline. Permeable rocks

A. Moura Bernardes (✉)
Programa de Pós Graduação em Engenharia de Minas, Universidade Federal do Rio Grande do Sul (UFRGS), Metalúrgica e de Materiais (PPGE3M), Porto Alegre-RS, Brazil
e-mail: amb@ufrgs.br

M. A. S. Rodrigues
Universidade FEEVALE, Novo Hamburgo, Brazil
e-mail: MarcoR@feevale.br

A. Moura Bernardes et al. (eds.), *Electrodialysis and Water Reuse*,
Topics in Mining, Metallurgy and Materials Engineering,
DOI: 10.1007/978-3-642-40249-4_6,

also form large underground reservoirs, making water available from both surface and groundwater sources. Wherever people work and live there will be an increase in toxic substances, nontoxic salts and pathogens entering the water cycle. Industrial contamination, although more localized, is often more serious. The extensive nature of agriculture makes it the most serious threat to drinking water quality, mainly due to the diffuse nature of such pollution, which makes it difficult to control. Any material or chemicals that find their way into the resource may need to be removed before supply [15].

6.2 Water Quality

Water quality requirements differ according to the application for which the water is required. While surface waters of the highest quality can generally be used for any freshwater need, many applications do not have such stringent requirements. The exact requirements for any particular end use will vary with local conditions and with government objectives. The setting of standards for this purpose is a complicated process, which requires consideration of such factors as toxicity, effects on biota, agricultural needs, risk of illness, and aesthetics (e.g., for recreational use). It also may require considering the effects of short- or long-term consumption and use, the availability and efficiency of processing methods to remove impurities, any seasonal variations in supply characteristics together with possible alternatives, and wastewater treatment methods [18].

Chemical water quality is determined by the quantity and diversity of organic and inorganic chemicals within it. Likewise, microbial water quality is dictated by the presence or absence of beneficial and pathogenic microorganisms [3]. Water rapidly absorbs both natural and man-made substances and, in general, this will make the water unsuitable for drinking prior to some form of treatment. The objective of water treatment is to produce an adequate and continuous supply of water that is chemically, bacteriologically and aesthetically pleasing. More specifically, water treatment must produce water that is [15]:

- palatable (i.e. no unpleasant taste);
- safe (i.e. does not contain pathogens or chemicals harmful to the consumer);
- clear (i.e. free from suspended solids and turbidity);
- colorless and odorless (i.e. aesthetic to drink);
- reasonably soft (i.e. allows consumers to wash themselves, their clothes and dishes without using excessive quantities of detergents or soap);
- non-corrosive (i.e. to protect pipework and prevent leaching of metals from tanks or pipes);
- low-organic content (i.e. high-organic content results in unwanted biological growth in pipes and storage tanks that often affects quality).

6.3 Conventional Drinking Water Treatment System

Many aquifers and isolated surface waters are high in water quality and may be pumped from the supply and transmission network directly to any number of end uses, including human consumption, irrigation, industrial processes, or fire control. However, clean water sources are the exception in many parts of the world, particularly regions where the population is dense or where there is heavy agricultural use. In these places, the water supply must receive varying degrees of treatment before distribution [25].

The most common water treatment process, known as conventional treatment, consists of disinfection, coagulation, flocculation, sedimentation, filtration, and disinfection (Fig. 6.1).

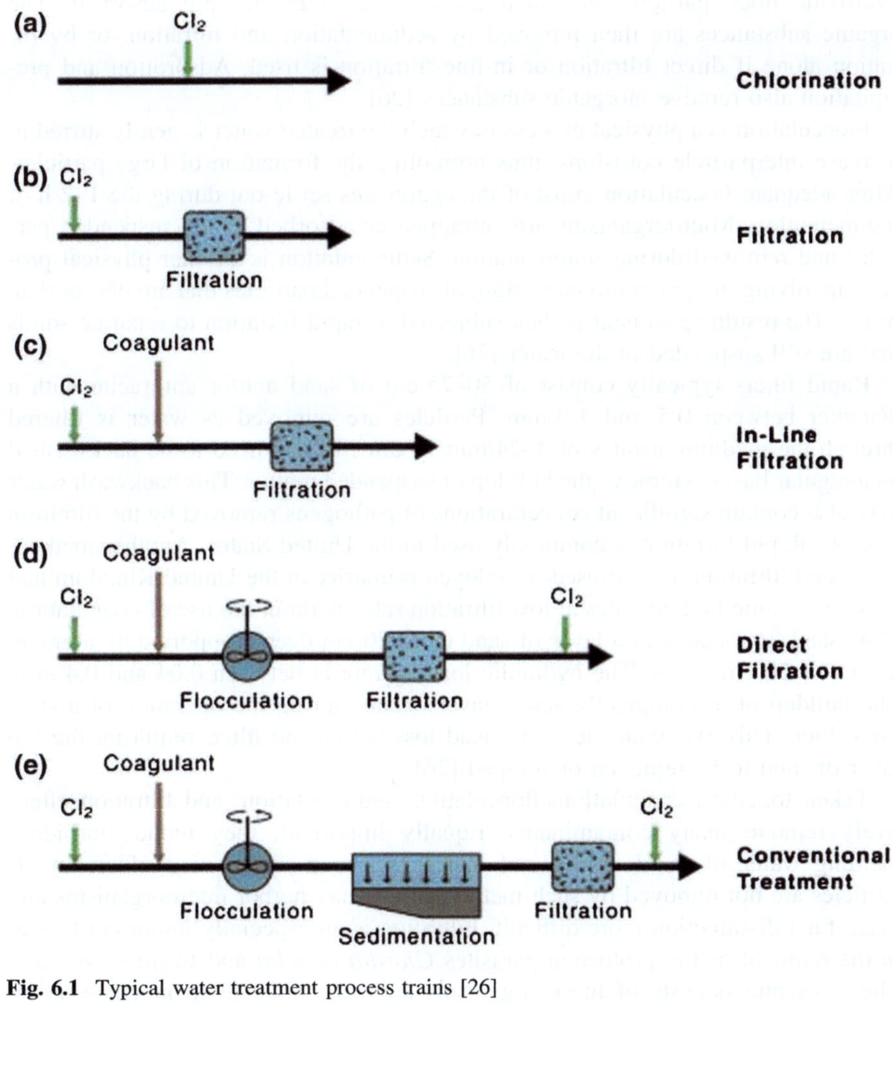

Fig. 6.1 Typical water treatment process trains [26]

Disinfection consists of a treatment process by chlorination. Coagulation involves the addition of chemicals to facilitate the removal of dissolved and suspended solids by sedimentation and filtration. The most common primary coagulants are hydrolyzing metal salts, most notably alum [$Al_2(SO_4)_3.14H_2O$], ferric sulfate [$Fe_2(SO_4)_3$], and ferric chloride [$FeCl_3$]. Additional chemicals that may be added to enhance coagulation are charged organic molecules called polyelectrolytes; these include high molecular weight polyacrylamides, polyamines, and starch. These chemicals ensure the aggregation of the suspended solids during the next treatment step, flocculation. Sometimes polyelectrolytes (usually polyacrylamides) are added after flocculation and sedimentation as an aid in the filtration step. Coagulation can also remove dissolved organic and inorganic compounds. Hydrolyzing metal salts added to the water may react with the organic matter to form a precipitate, or they may form aluminum hydroxide or ferric hydroxide flock particles in which the organic molecules are adsorbed. The organic substances are then removed by sedimentation and filtration, or by filtration alone if direct filtration or in-line filtration is used. Adsorption and precipitation also remove inorganic substances [26].

Flocculation is a physical process in which the treated water is gently stirred to increase interparticle collisions, thus promoting the formation of large particles. After adequate flocculation, most of the aggregates settle out during the 1–2 h of sedimentation. Microorganisms are entrapped or adsorbed in the suspended particles and removed during sedimentation. Sedimentation is another physical process, involving the gravitational settling of suspended particles that are denser than water. The resulting effluent is then subjected to rapid filtration to separate solids that are still suspended in the water [26].

Rapid filters typically consist of 50–75 cm of sand and/or anthracite with a diameter between 0.5 and 1.0 mm. Particles are removed as water is filtered through the medium at rates of 4–24/min/10 dm^2. Filters need to be backwashed on a regular basis to remove the buildup of suspended matter. This backwash water may also contain significant concentrations of pathogens removed by the filtration process. Rapid filtration is commonly used in the United States. Another method, slow sand filtration, is also used. Employed primarily in the United Kingdom and Europe, this method operates at low filtration rates without the use of coagulation. Slow sand filters contain a layer of sand (60–120 cm deep) supported by a gravel layer (30–50 cm deep). The hydraulic loading rate is between 0.04 and 0.4 m/h. The buildup of a biologically active layer occurs during the operation of a slow sand filter. This eventually leads to head loss across the filter, requiring the top layer of sand to be removed or scraped [26].

Taken together, coagulation, flocculation, sedimentation, and filtration effectively remove many contaminants. Equally important, they reduce turbidity, yielding water of good clarity and hence enhanced disinfection efficiency. If particles are not removed by such methods, they may harbor microorganisms and make final disinfection more difficult. Filtration is an especially important barrier in the removal of the protozoan parasites *Giardia lamblia* and *Cryptosporidium.* The cysts and oocysts of these organisms are very resistant to inactivation by disinfectants, so disinfection alone cannot be relied on to prevent waterborne illness. Because of their smaller size, however, viruses and bacteria can pass through the filtration process. Removal of viruses by filtration and coagulation depends on their attachment to particles (adsorption), which is dependent on the surface charge of the virus. This is related to the isoelectric point and is both strain and type dependent. The variations in surface properties have been used to explain why different types of viruses are removed with different efficiencies by coagulation and filtration. Thus, disinfection remains the ultimate barrier to these microorganisms [26].

Generally, disinfection is accomplished through the addition of an oxidant. Chlorine is by far the most common disinfectant used to treat drinking water, but other oxidants, such as chloramines, chlorine dioxide, and ozone, are also used. Although ultraviolet (UV) light can be used, it does not leave a residual and usually a secondary disinfectant (i.e., chlorine) is added [26].

In many cases an additional treatment is necessary to remove suspended, colloidal and dissolved constituents remaining after conventional treatment. Dissolved constituents may range from relatively simple inorganic ions, such as calcium, potassium, sulfate, nitrate and phosphate, to an ever increasing number of highly complex synthetic organic compounds [36].

Source protection, although not strictly in the category of treatment, can play a critical role in supporting the quality of water supplies. As of 1998, the city of New York spent approximately $1 billion to conserve and protect catchment areas in upper New York State [33].

6.4 Membrane Processes Applied to Drinking Water Production

The ability of a water treatment facility to deliver a wholesome and high quality product to its consumers is usually achieved by employing advanced treatment methods. Some of the more advanced technologies used today focus on the ability to use instrumentation to control the optimal quantities of chemicals needed in the operations of the process (i.e., coagulation, flocculation, sedimentation, and filtration). Such instrumentation is also used to continuously monitor water quality throughout the water treatment process. In addition to instrumental controls, numerous advanced treatment methods can be used to reduce both biological and chemical pollutants. In the United States, as well as in other countries, activated carbon or granulated activated carbon (GAC) has been used to remove organic chemicals from water since the early 1900s. Ion exchange resins have been employed to remove inorganic metals for decades. Ozonation has been successfully used in Europe, instead of chlorine and bromine, to both disinfect water and oxidize organic constituents. Ozonation in combination with hydrogen peroxide (i.e., advanced oxidation) has been used in the United States to remove persistent compounds such as methyl tertiary butyl ether (MTBE). Ultraviolet light, which has also been used in Europe, is being evaluated at a number of municipal treatment facilities in the United States for bacterial and viral control [33].

As far back as 1985, the French have made use of membrane separation technology (i.e., reverse osmosis) for the removal of organic and inorganic compounds. Today, a number of reverse osmosis installations are operating in the United States and other industrialized nations. Reverse osmosis membranes can hold back a wide range of micro-pollutants, including pesticides and pathogenic organisms, and can replace the need for both ozonation and activated carbon filtration. Nanofiltration, which operates at lower pressures than reverse osmosis, can remove compounds in the 300–1,000 molecular weight range [33].

6.5 Electrodialysis Applied to the Desalination and Purification of Brackish Water

Electrodialysis (ED) was commercially introduced during the 1960s well before the reverse osmosis for water desalination. The most important large-scale application of ED is the production of potable water from brackish water, in which ED competes directly with reverse osmosis (RO), nanofiltration, multistage flash evaporation, in addition to ion exchange treatments [5, 19, 32]. For certain feed water compositions, however, it has significant technical advantages over the competing processes [32].

There are several features that make electrodialysis especially suited for the desalination of brackish waters. Electrodialysis is less sensitive to membrane fouling and scaling than reverse osmosis. Therefore, higher recovery rates can be achieved and brine disposal problems can be minimized. Ion exchange membranes can be operated at elevated temperatures in excess of 50 °C without any effect on their chemical and mechanical stability or performance. The membranes are also quite stable at high and low pH-values and less sensitive to oxidizing agents than most reverse osmosis membranes. Brackish waters obtained from deep wells have often a high concentration of divalent ions and come at elevated temperature. These raw waters are difficult to desalt by reverse osmosis without significant pre-treatment, including cooling. For electrodialysis, however, they pose no problems and can be processed without any or a minimum of pre-treatment. Even if in some cases electrodialysis requires higher investments and operating costs than low pressure reverse osmosis, it is often the preferred process in brackish water desalination because of the clear technical advantages [32]. In addition, ED is generally the most economical process for water with relatively low salt concentrations (less than 5,000 ppm). Another significant feature of ED is that the salt can be concentrated to comparatively high values (in excess of 18–20 % by wt.) without affecting the economics of the process [19]. Furthermore, when ED is applied to brackish water desalination, a large fraction, typically 80–95 % of the brackish feed, is recovered as product water. The degree of water recovery is limited by precipitation of insoluble salts in the brine [5].

The desalination of brackish water by electrodialysis has some disadvantages compared to reverse osmosis desalination because neutral toxic components such as viruses or bacteria are not removed, and the product water may require a post-treatment procedure when used for the potable water supply. Furthermore, the generation of chlorine gas at the anode can lead to corrosion problems in the plant's surroundings if the venting is insufficient. Despite the fact that electrodialysis reversal is significantly less sensitive to membrane fouling than reverse osmosis, some pretreatment of the feed water is required to achieve trouble free operation. The iron and manganese concentrations, in particular, must be kept below 0.3–0.05 mgL^{-1}. The removal of iron and manganese is achieved by oxidation of the dissolved iron and manganese with potassium permanganate prior to precipitation and filtration [32].

The Electrodialysis process has been used commercially in the United States to desalt brackish waters for over 50 years. The first municipal installation in Coalinga, California, was commissioned in 1959. The first plant continuously used to produce the total water supply for a community was installed in Buckeye, Arizona, in 1962. Development of the Electrodialysis Reversal (EDR) process in the early 1970's resulted in a system featuring reduced operation and maintenance due to the elimination of the need to continuously feed acids and chemicals. This process is in use in typical municipal applications in Dell City, Texas, and Coupeville, Washington, along with industrial applications in electronics manufacture. Safe drinking water is produced by EDR units at a number of highway rest and service areas [14].

Worldwide more than 2,000 plants are installed for the desalination of brackish water [19]. A global installed capacity of 1–2 million m^3 per day seems to be a realistic estimation. The capacity of the individual desalination plants varies widely from a few 100 m^3/day to 20,000 m^3/day. Container mounted units with a capacity of 50–300 m^3/day are often used to supply potable water to isolated hotels, small islands or the labor force in camps in isolated desert areas. Larger installations with capacities between 10,000 and 30,000 m^3/day are used for municipal water supply or in industrial applications [32].

In India, the Central Salt & Marine Chemicals Research Institute (CSMCRI) installed several brackish water desalination plants of different capacities in rural areas during 1985–1995 [16, 19, 24]. ED units powered by photovoltaic cell were also developed and installed for providing potable water from brackish water [1, 19, 21]. A substantial number of installations of ED/EDR units can also be found in Russia and China for the production of potable water (Yuan [41]).

Over the last 10–15 years, numerous advances in membrane and system technology have made EDR an especially attractive technology, both in terms of performance and cost effectiveness. Many applications of EDR technology can be found worldwide and different applications are involved.

Table 6.1 shows a list of some of the industrial installations around the world. Different suppliers and membrane manufacturers are available. The biggest supplier is General Electric Water & Process (USA), which acquired Ionics Inc. in 2004.

Table 6.1 Some worldwide EDR systems [37]

Location	Country	Application		Production m^3/d	Year
Eurodia					
Montefano	Italy	Ground water	Nitrate removal	1,000	1991
Munchenbuschsee	Switzerland	Ground water	Nitrate removal	1,200	1996
Kleylehof	Austria	Ground water	Nitrate removal	3,500	1997
General electric water and process (Fomerly Ionics inc)					
Abrera, Bcn	Spain	Surface water	Bromide reduction	200,000	2008
Magna, Utah	USA	Ground water	Arsenic reduction	22,728	2008
Sherman, Texas	USA	Surface water	Salinity reduction	27,700	1993–1998
Suffolk, Virginia	USA	Ground water	Fluoride reduction	56,000	1990
Sarasota, Or	USA	Ground water	Hardness and salts reduction	45,420	1995
Maspalomas	Spain	Ground water	Salinity reduction	37,000	1986
Barranco Seco, Canary Islands	Spain	Waste water	Reuse	26,000	2002
Bermuda Waterworks	Bermudas	Ground water	Hardness and nitrate reduction	2,300	1989
Falconera, Valencia	Spain	Ground water	Nitrate reduction	16,000	2007

Ionics Inc. developed the first ED commercial system in the 1950s. Today, it is possible to use different suppliers of this technology, who either delivers the entire system, or only the membranes as internal suppliers. In this sense, it is possible to cite: MEGA (Czech Republic), Eurodia (France), Hidrodex (Brasil) and Tecnoimpanti (Italy) [37].

GE currently has an installed capacity of approximately 950,000 m^3/day of treated water by electrodialysis. Currently, some regions in the U.S., such as Oklahoma, Arizona, Sulfolk (Virginia), Texas and San Diego, as well as cities and regions in Europe, such as Barcelona and the Canary Islands, Spain, and Donnington in the UK, use the ED technique for the treatment of brackish or groundwater [13, 39].

The City of Suffolk (population of approximately 55,000) is located in Southeast Virginia, USA. In the late 1980s the City evaluated a number of source alternatives for expanding its water supply to meet the increasing demand for water. The well water has moderate levels of salts (689 $mg{\cdot}L^{-1}$). However, the fluoride level of 4.6 $mg{\cdot}L^{-1}$ is higher than the primary limit of 4.0 $mg{\cdot}L^{-1}$. In addition, the sodium concentration of 191 mg exceeds Suffolk's water quality guidelines of 50 $mg{\cdot}L^{-1}$ for potable water. Hence, a membrane desalination treatment process was required to raise the well water quality to meet drinking water standard. Ionics succeeded in the treatment of this feeding water using EDR. The EDR facility started operation at Suffolk in August of 1990. Since its commissioning, the plant has produced over five billion gallons of potable water [35].

Spain is one of the most arid countries in Europe and has implemented strategies for the desalination of brackish water for a long time. This effort resulted in the installation in 2009 of an electrodialysis reversal (EDR) plant near Barcelona, operated by the company Aigües Ter-Llobregat (ATLL). ATLL has a drinking water treatment station (DWTS), located in Abrera, which draws water directly from the river Llobregat. It presents a low and irregular flow and some quality problems, such as high salinity with a significant presence of sulphates, Ba^{+2}, Sr^{+2}, Na^{+}, Ca^{+2}, K^{+}, Cl^{-} and specially Br^{-}. Furthermore, many problems are associated with the increase in micropollutant and microbiological levels due to both urban and industrial sewage. Consequently, the high levels of bromide (ranging between 0.5 and1.2 mg/L) and natural organic matter produce high concentrations of trihalomethanes (THMs) after chlorination, showing a high brominated profile. To minimize the THM problem, ATLL searched for a new technology based on a membrane process. After many studies the final decision was the inclusion of a new EDR step after Granular Activated Carbon (GAC) filtration. This plant is the world's largest desalination plant using this technology, and a new example of a large scale application of a desalting technology to improve the quality of drinking water [37]. The electrodialysis plant treats works together with a conventional treatment plant. The desalted product of the EDR plant is mixed with the product of the conventional treatment plant to produce a combined stream which suits the drinking water needs of the region and is cost effective. The process operates with a flow rate of 2.4 m^3/s, a water recovery yield of 85–90 % and a reduction of water conductivity of 60–80 % [13]. The plant started operating on a trial basis in June 2008, and came into the normal operation in April 2009 [37].

6.6 Electrodialysis Aapplied to Nitrate Contaminated Potable Water

Nitrate contamination of drinking water is a widespread problem. It has long been known that levels of nitrates exceeding the 10 $mg.L^{-1}$ (as N-nitrogen) limit are associated with certain health problems. Although high nitrogen concentrations in drinking water are found mainly in regions of intensive agricultural use, there are source of nitrate contamination that are not agricultural. Fertilizer runoff, farm animal waste, and septic tank discharges all percolate through the soil into groundwater aquifers and, ultimately, into water supplies [37].

In Brazil, groundwaters are an important source of supply, and in the state of São Paulo alone over 20,000 deep wells and an immeasurable amount of shallow dug wells provide water for public supply, irrigation and industrial use [8]. In Natal (Northeast of Brazil), 65 % of the water distributed by the Company for Sanitation and Sewerage of Rio Grande do Norte derives from groundwater and the quality of these waters has deteriorated due to increasing urban activities. Among these activities, the infiltration of the soil with wastewater from septic tanks has the most severe impact. Thus, the groundwater from the city has high nitrate levels [7].

Different technologies are available to reduce the concentration of nitrates in drinking water, such as Biological Denitrification, Ion Exchange, Reverse Osmosis and Electrodialysis [9, 11].

The conventional treatment for removing nitrogen compounds in industrial wastewater is the biological treatment. This process, however, is greatly influenced by the temperature and organic load. For the biological treatment to remove nitrogen compounds of the groundwater for the public water supply to work, the addition of carbon-containing compounds is required because of the low concentration of organic material in these waters [9]. Thus, for natural waters the biological denitrification must be carried out with the addition of methanol or ethanol.

Other technologies have been evaluated for reducing nitrate concentrations in water, such as ion exchange with a strong anionic resin and regeneration with NaCl. This process has the disadvantage of adding chloride to the water and of not removing other dissolved solids in the form of cations [11].

In this context, the processes that apply membranes as a separating agent, such as reverse osmosis (RO) and electrodialysis (ED). Appear to be valid alternatives. These processes remove other ions in addition to nitrate, which results in decreased levels of sodium, chloride, hardness, etc. For waters with high salinity, this represents a large increase in the quality of the treated water [12]. For water with moderate concentrations of nitrate, the product of RO or ED can be mixed with the feed water to achieve the desired level of nitrates with higher levels of recovery and production [11].

Several authors have evaluated the use of electrodialysis processes for the removal of nitrates from drinking water in terms of process parameters and application conditions [10, 17, 20, 23, 28, 29, 30, 40].

Some authors have evaluated the use of electrodialysis for arid regions and with groundwater consumption. Banasiak et al. [6] evaluated the use of ED for treating water from a remote region in Australia. For this study, the groundwater of the Pine Hill Farm, located at 140 km from Alice Spring, containing a series of inorganic contaminants and a total concentration of dissolved solids of 5 $g \cdot L^{-1}$, was selected. The fluoride (2.8 $mg \cdot L^{-1}$) and nitrate (31.1 $mg \cdot L^{-1}$) concentrations exceed the water quality parameters of drinking water. The results showed that ED is capable of achieving water with drinking water quality levels.

Many electrodialysis plants have been recently installed specifically for the removal of nitrates from drinking water, showing that nitrates and nitrites are removed efficiently and economically using EDR [37]. In Israel, a GE plant (EDR 2020) was installed to reduce the levels of nitrate from water from 100 $mg \cdot L^{-1}$ to 45 $mg \cdot L^{-1}$, with 94 % water recovery. In the city of Kazusa, Japan, the technique was implemented to reduce nitrate levels from 80 mg·L^{-1} to 27 mg·L^{-1}. In Bermuda a plant removes 86 % of the nitrate concentration [13]. In Nagasaki, Japan, Astom corporation has installed an electrodialysis plant to remove nitrates, producing 125 m^3/day of drinking water [4].

EDR is also used since 2007 in Gandia, a tourist destination on the Spanish Mediterranean coast. The area sees peak demand during the summer months when the population almost triples, forcing the municipality to find alternate wells to feed the community. Upon analysis of the wells, it was determined that the nitrate levels were too high to meet the drinking water standard. The well samples had up to 80 mg·L^{-1} as nitrate (18.1 mg·L^{-1} as N). Treatment was necessary to produce acceptable levels of nitrate in the product water. An evaluation was conducted and EDR was selected as the technology of choice for the Gandia treatment plants. EDR offered high recovery while effectively reducing the nitrate levels to under 25 mg/L (5.6 mg/L as N). EDR was piloted on the wells to verify the nitrate removal and the operating cost estimates for power requirements and chemical consumption. The pilot study was successful, and the final systems were designed to achieve around 90 % water recovery with an overall removal of 73 % of the nitrate [31].

6.7 Concluding Remarks

Several studies comparing desalination technologies with electrodialysis were performed on groundwater, surface water and effluents [22, 27, 34, 38]. These studies led to favorable evaluation of electrodialysis because of several factors, such as: high recovery, simple pretreatment of the feed water, less susceptibility to fouling and deposits and lower operating costs.

The lifetime of membranes depends on the working conditions, but is often around 5 years. In Gandia, Valencia, Spain, a plant for the treatment of drinking water has been operating by electrodialysis reversal to remove nitrate since 2007. Its EDR system allows the desalination of water with very high recoveries, in the order of 90 %, with minimal pretreatment and lower operating costs (0.15–0.20 €/m^3). The plant has been in operation for 6 years, with the production of 8,000 m^3/day, without exchanging membranes. The concentrate produced is sent to the sewage treatment plant in the city [2].

References

1. Adiga MR, Adhikary SK, Narayanan PK et al (1987) Performance analysis of photovoltaic electrodialysis desalination plant at Tanote in Thar desert. Desalination 67:59–66. doi:10.1016/0011-9164(87)90232-3
2. Aguas De V (2012) http://www.aguasdevalencia.es/ Accessed 27 Oct 2012
3. Artiola J, Pepper I, Brusseau M (2004) Environmental monitoring and characterization. Academic Press, San Diego p 204

4. ASTOM Corporation (2010) http://www.astom-corp.jp/en/index.html. Accessed 10 Oct 2010
5. Baker RW (2012) Membrane technology and applications. Wiley & Sons Ltd., Chichester, p 575
6. Banasiak LJ, Kruttschnitt TW, Schäfer AI (2007) Desalination using electrodialysis as a function of voltage and salt concentration. Desalination 205(1–5):38–46. doi:10.1016/j.desal.2006.04.038
7. Cabral NMT, Righetto AM, Queiroz MA (2009) Comportamento do nitrato em poços do aqüífero Dunas/Barreiras nas explotações Dunas e Planalto, Natal, RN, Brasil. Eng Sanit Ambient 14(3):299–306
8. CETESB (2010) http://www.cetesb.sp.gov.br/Agua/rios/informacoes.asp. Accessed 18 July 2012
9. Della Rocca C, Belgiorno V, Meriç S (2007) Overview of in situ applicable nitrate removal processes. Desalination 204(1–3):46–62. doi:10.1016/j.desal.2006.04.023
10. Elmidaoui A, Menkouchi-Sahli MA, Tahaikt M et al (2003) Selective nitrate removal by coupling electrodialysis and a bioreactor. Desalination 153(1–3):389–397. doi:10.1016/S0011-9164(02)01133-5
11. Elyanow D, Persechino, J (2010) Advances. In: Nitrate removal. Technical paper. Ge water and process technologies. Technical paper. GE water and proocess technologies. Available from http://www.gewater.com. Accessed 10 July 2010
12. Garcia F, Ciceron D, Saboni A et al (2006) Nitrate íons elimination from drinking water by nanofiltration. Sep Purif Technol 52(1):196–200. doi:10.1016/j.seppur.2006.03.023
13. GE Energy (2010): GE helps catalonia region make the most of scarce water supplies, http://www.gepower.com/about/press/en/2009_press/051909.htm Accessed 10 Oct 2012
14. Goldstein AL (1979) Electrodialysis on the American continent. Desalination 30(1):49–58. doi:10.1016/S0011-9164(00)88432-5
15. Gray NF (2005) Water technology. Elsevier Butterworth, Heinemann. p 645
16. Harkare WP, Adhikary SK, Narayanan PK et al (1982) Desalination of brackish water by electrodialysis. Desalination 42(1):97–105. doi:10.1016/S0011-9164(00)88745-7
17. Hell F, Lahnsteiner J, Frischherz H et al (1998) Experience with full-scale electrodialysis for nitrate and hardness removal. Desalination 117(1–3):173–180. doi:10.1016/S0011-9164(98)00088-5
18. Hocking MB (2005) Raw water processing and wastewater treatment. In: Handbook of chemical technology and pollution control 3rd edn. Academic Press, San Diego, pp 139–174
19. Kariduraganavar MY, Kittur AA, Kulkarni SS (2012) Ion exchange membranes: preparation, properties, and applications. In: Inamuddin L, Mammad L (eds.) Ion exchange technology I: theory and materials. Springer, New York, pp 233–276. doi: 10.1007/978-94-007-1700-8
20. Kesore K, Janowski F, Shaposhnik VA (1997) Highly effective electrodialysis for selective elimination of nitrates from drinking water. J Membr Sci 127(1):17–24. doi:10.1016/S0376-7388(96)00282-7
21. Kuroda O, Takahashi S, Kubota S et al (1987) An electrodialysis sea water desalination system powered by photovoltaic cells. Desalination 67:33–41. doi:10.1016/0011-9164(87)90229-3
22. Lozier CJ, Smith G, Chapman JW et al (1992) Selection, design, and procurement of a demineralization system for a surface water treatment plant. Desalination 88:3–31. doi:10.1016/0011-9164(92)80103-G
23. Midaoui AE, Elhannouni F, Taky M et al (2002) Optimization of nitrate removal operation from ground water by electrodialysis. Sep Purif Technol 29(3):235–244. doi:10.1016/S1383-5866(02)00092-8
24. Narayanan PK, Harkare WP, Adhikary SK et al (1985) Performance of an electrodialysis desalination plant in rural area. Desalination 54:145–150. doi:10.1016/0011-9164(85)80013-8
25. Peirce JJ, Weiner RF, Vesilind PA (1998) Environmental pollution and control 4th edn. Elsevier, Boston, p 392
26. Gerba CP (2009) Drinking water treatment. In: Maier R, Pepper I, Gerba CS (2009) Environmental microbiology 2nd edn. Elsevier, Boston pp 532–538, 624

27. Ryabtsev AD, Kotsupalo NP, Titarenko VI et al (2001) Set-up involving electrodialysis for production of drinking-quality water from artesian waters with salt content up to 8 kg/m^3 with productivity up to 1 m^3/h. Desalination 136:333–336. doi:10.1016/S0011-9164(01)00196-5
28. Sahli MAM, Annouar S, Mountadar M et al (2008) Nitrate removal of brackish underground water by chemical adsorption and by electrodialysis. Desalination 227(1–3):327–333. doi:10.1016/j.desal.2007.07.021
29. Salem K, Sandeaux J, Molenát J et al (1995) Elimination of nitrate from drinking water by electrochemical membrane processes. Desalination 101(2):123–131. doi:10.1016/0011-9164(95)00015-T
30. Sata T (2000) Studies on anion exchange membranes having permselectivity for specific anions in electrodialysis—effect of hydrophilicity of anion exchange membranes on permselectivity of anions. J Membr Sci 167(1):1–31. doi:10.1016/S0376-7388(99)00277-X
31. Seidel C, Gorman C, Darby JL et al (2011) An assessment of the state of nitrate treatment alternatives. final report. the American water works association, inorganic contaminant research and inorganic water quality joint project committees, California, 136p. http://smallwatersystemsucdavis.edu/documents/JACOBS-UCDavisNitrateTECReportFINAL062211.pdf. Accessed 10 Dec 2012
32. Strathmann H (2004) Ion exchange membrane separation processes, vol 9. Elsevier, Amsterdam, pp 1–348
33. Sullivan PJ, Agardy FJ, Clark JJJ (2005) Water protection. The environmental science of drinking water. Elsevier, Burlington, pp 89–141
34. Swami MSR, Muruganandam L, Mohan V (1996) Recycle of treated refinery effluents using electrodialysis-a case study. Indian J Environ Prot 16(4):282–285
35. Tanaka Y (2007) Ion exchange membranes: fundamentals and applications, vol 12. Elsevier, pp 1–531
36. Tchobanoglous G, Burton FL, David-Stensel H (2003) In: Metcalf & Eddy, Inc. wastewater engineering: treatment and reuse 4th edn. McGraw-Hill Companies, New York p 1819
37. Valero F, Barceló A, Arbós R (2011) Electrodialysis technology: theory and applications. In: Scorr M (ed) Desalination, trends and technologies, InTech, pp 3–20. doi: 10.5772/14297. Available from http://www.intechopen.com/books/desalination-trends-and-technologies/electrodialysis-technology-theory-and-applications. Accessed 05 Nov 2012
38. Van der Hoek JP, Rijnbende DO, Lokin CJA et al (1998) Electrodialysis as an alternative for reverse osmosis in an integrated membrane system. Desalination 117(1–3):159–172. doi:10.1016/S0011-9164(98)00086-1
39. Werner TE, Von Gottberg AJM (1998) Five billion gallons later–operating experience at city of Suffolk EDR plant, the American desalting association. In: North American biennial conference and exposition, Williamsburg, 2–6 August 1998. Available from http://www.gewater.com/pdf/Technical%20Papers_Cust/Americas/English/TP1031EN.pdf
40. Yeon K-H, Song J-H, Shim J et al (2007) Integrating electrochemical processes with electrodialysis reversal and electro-oxidation to minimize COD and T-N at wastewater treatment facilities of power plants. Desalination 202(1–3):400–410. doi:10.1016/j.desal.2005.12.080
41. Yuan Z, Richard S, Tol J (2004) Implications of desalination for water resources in China—an economic perspective. Desalination 164(3):225–240. doi:10.1016/S0011-9164(04)00191-2

Chapter 7
Electrodialysis Treatment of Refinery Wastewater

Mara de Barros Machado and Vânia M. J. Santiago

Abstract Water is fundamental in the oil refining process, with on average 0.9 m^3 water being consumed per m^3 of oil processed. Water is used in the oil refining process for steam generation, cooling systems, potabilization, fire fighting, the removal of contaminants and services in general. Industries are having difficulties in obtaining water due to the growing scarcity and pollution of water, with increasing treatment costs as a consequence, which limits increases in productivity. It has therefore become necessary that industries implant a water management system, including its rationalization with process modification, choosing the best treatment method for discharge or reuse. For many industries, management systems and water reuse may be an important factor contributing to their survival. The wastewater generated in refineries is sent to a treatment plant, to reach the necessary quality for discharge. The composition of the wastewater varies greatly and depends on the complexity of the refining process. The reuse requires additional treatment processes and the selection of the method will depend on the characteristics of the wastewater and on the required water quality. Wastewater treatment processes using membranes are being increasingly used because of their high efficiency, small installation area, low consumption of chemicals and increased automation. Among the different membrane processes, electrodialysis has been increasingly employed in the industry for partial desalination aiming water reuse. In refineries, the electrodialysis reversal process is not yet widely used, but it has great growth potential. The evaluation of the pilots and industrial units in operation have shown the advantages over similar desalination processes, such as greater robustness, operational reliability and simplicity and less necessity of wastewater pre-treatment. This chapter will present an overview of the use of electrodialysis for water reuse in refineries.

M. de Barros Machado (✉) · V. M. J. Santiago
Petróleo Brasileiro S.A, Rio de Janeiro, Brazil
e-mail: mara_bm@petrobras.com.br

V. M. J. Santiago
e-mail: vaniamjs@petrobras.com.br

A. Moura Bernardes et al. (eds.), *Electrodialysis and Water Reuse*,
Topics in Mining, Metallurgy and Materials Engineering,
DOI: 10.1007/978-3-642-40249-4_7, © Springer-Verlag Berlin Heidelberg 2014

7.1 Water Use and Wastewater Generation in Refineries

Petroleum is a complex combination of organic and inorganic compounds, mostly hydrocarbons (aliphatic, alicyclic and aromatic). To generate quality products with commercial value, the petroleum is submitted to refining processes, where it is separated into fractions and subjected to quality improvement treatments. The combination of physicochemical characteristics of the petroleum being processed, the demand for derivative products in the region in question and the availability of technologies will lead to the ideal process combination for each refinery.

Water in a refinery is used for cooling systems, steam generation, chemical processes (petroleum desalination, dissolution of salts, preparation of chemical products), flushing equipment, pipelines, hydrotests, washing products, irrigation, potable water and fire fighting. Cooling systems account for up to 90 % of the total water requirements in refineries and their demand is estimated as a function of the complexity of the refinery processes [15].

In refineries, an average of 0.9 m^3 of water is used per m^3 of petroleum processed, generating 0.3 m^3 of wastewater. Wastewater includes condensed steam, stripping water, residual caustic solutions, cooling tower and boiler blowdown, wash water, alkaline and acid waste neutralization water, oil tank drainage, wastewater from oil desalters and other processes associated with water [14]. The characteristics of the wastewater will depend on the raw materials processed, the processing steps used and the incorporation of such substances as solvents, additives etc. The main contaminants of the wastewater are free and emulsified oils, phenols, mercaptans, sulfides, ammonia, cyanide, dissolved solids and organic compounds.

7.2 Conventional Wastewater Treatment in Refineries

The wastewater of refineries undergoes preliminary, primary and secondary treatment. Generally, after the secondary treatment, the water reaches the quality required for discharge. In the case of water reuse, and depending on the application, the removal of specific compounds is important and a tertiary treatment will be necessary.

7.2.1 Preliminary Treatment

In the preliminary treatment, coarse solids and substances that may affect the treatment and the equipment used are removed. In order to accomplish this goal, different unit operations are applied.

The first operation is the segregation of the contaminated water stream, oily water, domestic sewage and rainwater drainage streams.

For water to be classified as contaminated, it usually has to be originated from processes that may or may not contain oil at some point: the blowdown from cooling towers, pumps and unit process areas, pipelines, tanks, compressors areas, and loading truck areas.

The oily water stream is composed of waters from pumps and unit areas, loading and unloading areas for trucks and wagons, solid waste disposal and/or treatment areas (landfarming), service stations, garages, fire fighting training areas, produced water (water that is found in reservoirs and is produced with the oil), oil tank drainage, sour water, desalter wastewater and ballast water.

The domestic sewage stream is composed of water from sinks, showers, toilets, kitchens, water fountains etc.

The rainwater stream collects water from the administrative areas, buildings, streets, liquefied petroleum gas tanks and cylinders, and other areas that are not subject to contamination.

After segregation, a grating operation is performed that serves to retain the coarse solids that may come along with the wastewater. These solids may damage the subsequent equipment and reduce the treatment efficiency. It can be simple (manual cleaning) or mechanical (mechanical cleaning).

At the desanding stage, the sand and the settleable solids are removed, as a way to protect the subsequent equipments. It is installed upstream of the primary treatment. The main parameters in the design of desanding units are flow, wastewater velocity and minimum residence time for settling.

Afterwards, the equalization basin should homogenize and absorb the wastewater quality and flow variations in order to avoid the overload of the primary treatment unit.

7.2.2 Primary Treatment

The first step in the primary treatment process is to remove the oil from the wastewater. Refinery wastewater contains free and emulsified oils. The removal of free oil is done by the gravimetric process. The flotation step is performed in order to remove emulsified oil. During both processes a coagulant is often—dosed in the feed water (such as ferric chloride or aluminum sulfate) to flocculate the suspended matter and increase the efficiency of the process. In some cases the effluent from flotation goes straight to the secondary treatment. Depending on the wastewater characteristics, however, sometimes a coagulation/floculation/sedimentation operation is also applied. After the primary treatment, another equalization step is applied.

7.2.3 Secondary Treatment

The secondary process consists of removing or reducing suspended and dissolved organic matter from the wastewater. Usually, the wastewater is submitted to biological treatment in this step, since this is the most effective type of treatment. It can remove up to 90 % of the organic matter in wastewater. The most important parameters removed in this process are the BOD (biochemical oxygen demand) and ammonia. Secondary treatment may require a separation process to remove the micro-organisms from the treated water prior to discharge or tertiary treatment.

7.3 Electrodialysis to Process Water Production

Electrodialysis in refineries can be applied on water, wastewater treatment and industrial water production.

The largest consumer of water in a refinery is the cooling system [15]. The cooling tower removes heat directly from the water through an air flow. A portion of the water passing through the tower is vaporized; solids and contaminants contained in the other part are then concentrated. To optimize costs and efficiency, water is recirculated several times in a tower, and a portion is discharged (blowdown). This discharge is necessary to control the increase in the concentration of solids and contaminants in the water, preventing the formation of deposits and corrosion of the surface of heat exchangers, which may reduce the efficiency of heat transfer and cause contamination of the product to be cooled [5].

Since 1992 the U.S. Environmental Protection Agency [7] recommends standards for water recirculation cooling systems (Table 7.1).

Water quality for steam generation is higher compared to cooling systems. In this case the electrodialysis must be associated to another process, for efficiency improvement in salinity reducing and for the removal of non ionic contaminants as, for example, silica. Generally the adsorption on ion exchange resins is done after electrodialysis. Table 7.2 shows the recommended standards for steam generation [8].

The use of ED/EDR has increased with the recent development of membranes and associated processes.

The successful application of electrodialysis for the treatment of cooling tower blowdown has been reported [2, 3, 19].

Examples of EDR application in refineries can be seen at the Oil Industry of Livorno, Italy, which treats the wastewater for reuse in make-up cooling systems and steam generation [1, 9]. Wastewater treatment for cooling system make up is also applied at a refinery of Petrobras, Brazil [13].

Another possible application of electrodialysis is the concentration of the saline waste of reverse osmosis. Due to limit of selectivity and the osmotic pressure of the concentrated solution, the concentrated stream of the osmosis can not exceed

Table 7.1 Recommended water quality for cooling system [7]

Parameter[a]	Recommended limit[b]
Chloride	500
Dissolved solids	500
Hardness	650
Alkalinity	350
pH[c]	6.9–9.0
Chemical oxygen demand	75
Suspended solids	100
Turbidity[c]	50
Biochemical oxygen demand[c]	25
Organics[d]	1
Ammonia (NH_4-N)[c]	1
Phosphate[c]	4
Silica	50
Aluminium	0.1
Iron	0.5
Manganese	0.5
Calcium	50
Magnesium	0.5
Bicarbonate	24
Sulfate	200

[a] All values in $mg{\cdot}L^{-1}$ except pH
[b] Water Pollution Control Federation (1989) Water reuse manual of practice, 2nd edn. Water Pollution Control Federation, Alexandria, Virginia, *apud* EPA (1992)
[c] Goldstein DJ, Wei I, Hicks RE (1979) Reuse of municipal wastewater as make-up to circulating cooling systems. In: proceedings of the water reuse symposium, vol 1. AWWA Research Foundation, Denver, Colorado *apud* EPA (1992)
[d] Methlyene blue active substances

certain values. In general, electrodialysis may be used when the discharge of large volumes of saline waste is restricted and/or a higher concentration is desirable, [17].

7.4 Comparison of Electrodialysis Reversal and Reverse Osmosis

Membrane desalination has been used for years in refineries. For industrial water generation, particularly steam generation, reverse osmosis is used.

The ease, in which membranes can be obtained, associated with the simple construction of the units, gave reverse osmosis a competitive advantage and market dominance since the 70's [16]. Nevertheless, in recent years electrodialysis reversal is being increasingly used in applications previously associated with reverse osmosis. Today, with the increasing demand for water reuse, a practical

Table 7.2 Recommended boiler water limits—Source: Boiler Water Quality Requirements and Associated Steam Quality for Industrial/Commercial and Institutional Boilers (American Boiler Manufacturers Association 2005 *apud* EPA [8])

Drum operating pressure (psig)	0–300	301–450	451–600	601–750	751–900	901–1000	1001–1500	1501–2000	OTSG
Steam									
TDS max (ppm)	0.2–1.0	0.2–1.0	0.2–1.0	0.1–0.5	0.1–0.5	0.1–0.5	0.1	0.1	0.05
Boiler water									
TDS max (ppm)	700–3500	600–3000	500–2500	200–1000	150–750	125–625	100	50	0.05
Alkalinity max (ppm)	350	300	250	200	150	100	n/a	n/a	n/a
TSS max (ppm)	15	10	8	3	2	1	1	n/a	n/a
Conductivity max (μmho/cm)	1100–5400	900–4600	800–3800	300–1500	200–1200	200–1000	150	80	0.15–0.25
Silica max (ppm SiO_2)	150	90	40	30	20	8	2	1	0.02
Feed water (condensate and makeup, after deaerator)									
Dissolved oxygen (ppm O_2)	0.007	0.007	0.007	0.007	0.007	0.007	0.007	0.007	n/a
Total Iron (ppm Fe)	0.1	0.05	0.03	0.025	0.02	0.02	0.01	0.01	0.01
Total Copper (ppm Cu)	0.05	0.025	0.02	0.02	0.015	0.01	0.01	0.01	0.002
Total hardness (ppm $CaCO_3$)	0.3	0.3	0.2	0.2	0.1	0.05	ND	ND	ND
pH (25° C)	8.3–10.0	8.3–10.0	8.3–10.0	8.3–10.0	8.3–10.0	8.8–9.6	8.8–9.6	8.8–9.6	n/a
Nonvolatile TOC (ppm C)	1	1	0.5	0.5	0.5	0.2	0.2	0.2	ND
Oily matter (ppm)	1	1	0.5	0.5	0.5	0.2	0.2	0.2	ND

assessment of both methods is required, guiding the best choice of treatment for each application.

Some factors need to be considered in evaluating the implementation of the two methods:

- Produced water quality.
- Pre-treatment required of the feed water.
- Consumption of chemicals.
- Lifetime of membranes.
- Energy consumption.

The pre-treatment systems are required to protect the membrane, but the degree of the treatment is quite different for the two technologies.

Salts with lower solubility affect both technologies, especially the calcium salts, in addition to organic contaminants, which must be removed by biological treatment, chlorination and/or activated carbon adsorption. Electrodialysis reversal is more tolerant to iron, admitting values of up to 0.5 ppm, while in osmosis this value should be <0.05 ppm. In osmosis, pre-treatment also requires the removal of manganese, aluminum, zinc and dechlorination. EDR membranes are more resistant to disinfectants, such as hypochlorite, chloramine and peroxides [17].

The electrodialysis reversal system, which aims the continuous production of demineralized water without the constant addition of chemicals during normal operation, uses reverse polarity to continuously control deposition and fouling.

In these systems the polarity of the electrodes is reversed 3–4 times every hour, changing the direction of the movement of ions inside the membrane module and thus controlling the fouling and scaling. In a typical system, reversibility occurs every 15 min and is done automatically.

Because of the reversibility, none of the stack channels are exposed to solutions of high concentrations for a period >15–20 min. Any initial precipitation of salt is carried and dissolved when the cycle is reversed.

EDR reduces the total organic carbon by 20–40 %, when ionizable organic matter of low molecular weight is present.

For drinking water production, there are units operating with feed water with silica concentration >140 ppm. Since ED does not remove nonpolar substances, silica concentrations are not a critical factor [26].

Iron, manganese and hydrogen sulfide in excess of 0.3 ppm is permitted if there is no air intake in the feed water container.

To improve the performance of the EDR on the treatment of critical wastewaters, chemicals can be dosed, usually acids or antiscalants are fed in the concentrate, reducing the precipitation of salts.

Problems with bacterial contamination have been reported as one of the main causes of membrane fouling in osmosis. This is a consequence of the characteristic of the membrane, which serves as a barrier between the feed water and the product, removing not only dissolved solids, but also bacteria, viruses, dissolved and insoluble substances. The pretreatment required for reverse osmosis is more demanding, representing up to 50 % of the total cost.

A disadvantage of electrodialysis is that it removes only ions. Unloaded or slightly loaded compounds are not removed, such as silica and boron. In addition, it does not form a barrier to organic compounds and microorganisms that change odor and taste and it therefore produces lower quality water when compared to osmosis.

The greatest advantage of electrodialysis is the membrane resistance, ensuring a longer life time than osmosis membranes, which range from 1 to 3 years. With regard to energy consumption, the electrodialysis reversal process has a lower energy consumption when applied to water with concentrations of dissolved solids that are below 2,000 ppm. Reverse osmosis is most advantageous when applied to water with salinity higher than 4,000 ppm [2, 3, 26]. For some authors electrodialysis is more economical than osmosis in higher concentrations of up to 8,000 ppm if maintenance costs are taken into account, and this limit is not due to inherent characteristics of the technology, but due to the necessity of optimization when when working with more saline waters [20–22].

Several studies comparing desalination technologies with electrodialysis were performed on groundwater, surface water and wastewater, including refineries. These studies led to the selection of electrodialysis because of several factors, such as: higher recovery, simple pretreatment of the feed water, less susceptibility to fouling and deposits and lower operating costs [12, 18, 23–25].

Electrodialysis should always be considered as an alternative to osmosis if the application can be efficiently serviced by both. In many cases, the advantages of electrodialysis can overcome its disadvantages in any specific factor and an analysis should be made that takes into account investment, operation and maintenance costs and the environmental impact.

7.5 Water Pretreatment for Electrodialysis Reversal

Pretreatment for electrodialysis requires the partial removal of compounds that can cause scaling and biofouling of membranes.

The cost of the pretreatment of the feed water varies greatly with its quality and is related to the salt concentration permitted in the EDR concentrate.

The effluents from refineries are subject to variations in composition, even over the course of a day. The wastewater treatment process must therefore have operational flexibility and robustness in order to minimize shutdowns because of fouling and scaling.

After the conventional treatment of effluents, the treated wastewater should pass through another treatment operation before it is submitted to EDR. The first step of this pretreatment is the removal of suspended solids and turbidity. The processes that may be used are micro or ultrafiltration with membranes or physicochemical processes such as coagulation, flocculation and filtration.

The colloidal material in most wastewaters has a negative charge and deposits on anionic membranes, inactivating the ionic change sites. The usual method for removing colloidal material is the pretreatment by coagulation-flocculation and filtration. In this step some chemicals are used to assist in the removal of solids and turbidity, such as coagulants, flocculants, acids or bases to adjust pH and oxidizing agent to degrade the organic matter. When membranes are used to remove solids, a preliminary step of coagulation may be needed to increase the effectiveness of colloid removal and/or oxidation [2, 3].

It is important to meet the limit concentrations for solids and turbidity to avoid problems in the subsequent stages of treatment and prevent clogging of the cartridge filter at the entrance of the electrodialysis unit.

The chemical products used in the pretreatment should be controlled. Cationic polymers used as coagulant, anionic polymers used as flocculants and dispersants are organics with high molecular weight and with charge, and may (when dosed in excess) cause fouling of the membrane surface [1].

When the concentration of iron in the feed water exceeds 0.3 ppm, its removal is recommended. Iron and manganese can be removed with potassium permanganate or by coagulation/flocculation. Filtration removes insoluble hydroxides of iron and manganese.

Organic compounds that were not degraded in biological treatment are removed by activated carbon filters and/or advanced oxidation processes (chloramines, ozone, hydrogen peroxide, chlorine dioxide…). In this step it is important to control the generation of fine carbon, and if necessary, to adopt operational changes in order to prevent clogging of the cartridge filter and passage into the membrane modules of the electrodialysis unit. When oxidative processes are used, the residual oxidant concentration must be controlled, not permitting it to exceed the limits for the feed water.

Hydrogen sulfide, if necessary, is removed by oxidation or aeration and filtration.

In most applications it is necessary to control microbiological growth, which can be accomplished through chlorination. The chlorine residual must be monitored and controlled to meet the limit values for EDR feed water. The limit value of chlorine must be informed by the supplier of the technology and will depend on the material of the membrane. In general, up to 0.5 mg/L of residual chlorine is tolerated in the feed water.

Cartridge filters are used upstream of electrodialysis units as the last protection against suspended solids. The nominal specification is usually 10–20 μm, with 10 μm being more widely used.

Scaling on the membrane surface due to the precipitation of salts, such as calcium sulfate, calcium carbonate, barium sulfate and magnesium hydroxide, tends to occur in the concentrate. This can be prevented by dosing an acid, generally hydrochloric acid, in the concentrate, not allowing the Langelier index to exceed 2.5.

Products that inhibit scaling can be used in the concentrate channel [6].

An EDR system can work with a saturation concentration of calcium sulfate of up to 175 % without problems. With chemical dosing (scale inhibitors) in the concentrated stream, this value increases to 300 %. There are studies of pilot plant operating successfully with 400 % of saturation.

Some optimization studies for the pretreatment of water for electrodialysis have indicated that the best operational results are achieved by adding a coagulation step upstream of the solid removal process and by dosing sodium hypochlorite for the control of microbiological growth [4, 22].

7.6 Energy Consumption

ED/EDR uses electricity to transfer ions through the stack and pumping solutions. Two or sometimes three pumping stages are required in the process [2, 3, 10, 25, 26].

Energy is also consumed by the reactions occurring at the electrodes and by the measuring instruments of operational processes. This consumption is often ignored or considered fixed between 1 and 3 % of the ionic transfer and pumping energy.

The energy cost for desalination depends on the feed water salinity, the amount of salt to be removed and the potential difference along the stack.

The average consumption in the desalination of brackish water ranges from:

- 0.5 to 1.1 kWh/m^3 of product for pumping
- 0.7 kWh/m^3 of product to each 1,000 mg TDS removed
- 5 % loss of total energy consumed

In a 2-year study with an electrodialysis reversal pilot unit that was used to treat the wastewater of a refinery in order for it to be reused, the energy consumption was 0.5 kWh/m^3 of water produced for desalination and 2.8 kWh/m^3 of water produced for pumping (consumption of two pumps), meaning a total of 3.3 kWh/m^3 produced water. The pilot plant was operated under the following conditions: 2 stages, voltage of 275 V and a flow rate of 1,100 L/h of produced water [13].

The energy consumption is affected by temperature. Higher temperatures increase the degree of ionization and mobility, reducing the electrical resistance of the solution and energy consumption. Another advantage is the reduction of solution viscosity, which results in lower costs. Some studies indicate an increase in the current density range of 1–4 % for every degree of rise in temperature [11]. Another study has indicated that the energy required can be reduced by 60–70 % when the temperature increases from room temperature to 70 °C [2]. As a rule, a 1 % reduction in energy consumption occurs for each increment of 0.5 °C at temperatures above 21 °C and a 1 % increase for every decline in temperature below 21 °C. One of the disadvantages of elevating the temperature is the deterioration of the membranes and spacers, because of the degradation of the polymeric material, in addition to an increase in the tendency of some compounds with lower solubility to precipitate. The maximum operating temperature for electrodialysis with high temperature projects is between 50 and 70 °C, limited by polymer degradation. In conventional ED/EDR designs, 45 °C is the limit due to the loss of rigidity of the spacers made of low density polyethylene. In practice, the membranes that are currently available, are limited to work until 38 °C as a high temperature threshold and 10 °C a low temperature threshold.

The higher the current density, the lower the membrane area required for a given rate of desalting, decreasing the investment costs and the rate of membrane replacement. On the other hand, energy costs are higher because the voltage increases in proportion to the current density. For each system there is an optimal current density in which the total cost reaches a minimum.

7.7 Operational Control and Maintenance

Operational control of the electrodialysis process aims to maintain the efficiency without operational problems, such as the fouling and scaling of membranes.

Operating conditions should be monitored on a daily basis: leaks, noises, operating pumps, valves, tank levels, flows, data from instruments installed at the plant, such as online analyzers; feed water, products and concentrate composition, preventive and operational procedures.

Depending on the EDR configuration, the electrode channel is separated from the product and concentrate channel. Thus it is possible to clean the electrode separately by an ECIP (Electrode Clean in Place) operation. This cleaning is periodic and automated, using an acid solution, usually hydrochloric acid, on the negative electrode (cathode), where reduction reactions occur with possible increase in pH and hence precipitation of salts.

The cleaning of the membranes, CIP (Clean in Place), should be done regularly to remove deposits and/or fouling of the membrane surface. For each application a cleaning protocol is provided, based on prevention or on the reduction of the desalination efficiency.

The reverse polarity system has a cleaning effect, since it reverses the conditions of the compartments, removing newly formed fouling films on the membrane surface, which enables longer intervals between one chemical cleaning operation and another.

Usually a 5 % solution (volume) of commercial hydrochloric acid is circulated in the stack for 30–60 min to remove inorganic deposits and precipitations.

Organic contaminants are removed in the following step with 5 % of NaCl solution (weight) with alkaline pH (adjusted with NaOH). The pH must not exceed the limit of the membrane material.

Small anions, enough to permeate the membrane, but with low electromobility, are retained on the membrane. These compounds, such as detergents, are difficult to remove.

When there are residual concentrations of polyelectrolytes and antifouling agents, a cleaning with chlorine can be performed. The maximum allowed residual chlorine concentration must be informed by the manufacturer of the membrane.

For more severe deposits, the stack must be disassembled and the membranes cleaned manually. The necessary infrastructure for such a process must be considered [2, 3, 26].

For long periods without operation, the membrane stacks must be preserved, filling the modules with a biocide solution. E.g.: an iodine solution.

Throughout the operation of the electrodialysis unit, the formation of precipitates outside of the stacks should be verified. The accumulation of precipitates outside the modules may lead to the forming of an electric bridge and the burning of the membranes.

To avoid damage to the membranes the external part of the stacks should be periodically washed with clean water.

Measuring the resistivity using a multimeter is a procedure for checking the integrity of membranes and electrodes. High resistivity results indicate problems and the stacks may have to be disassembled to evaluate the membrane and/or electrode.

7.8 Concluding Remarks

The reuse of wastewater in refineries produces environmental and social benefits, such as a reduction in water intake and wastewater discharge, benefits related to minimizing water consume and obtaining environmental operating licenses, a contribution to the company's environmental excellence, an improvement of the corporate image and an increase in market value.

Electrodialysis has been evaluated and validated as a desalination process that can be applied to the wastewater of refineries to reuse the water.

For some uses, electrodialysis has shown advantages over the reverse osmosis process, such as a greater operational robustness and lower energy consumption.

Other applications have been evaluated with good results and are technically feasible, such as the use in the cooling tower blowdown, the reuse as water makeup, the pretreatment of the reverse osmosis feed water, allowing greater operational continuity when using critical wastewaters, as well as on the treatment of the brine concentrate generated in desalination plants, in order to reduce the volume of waste.

Generally, electrodialysis in refineries is used after the biological treatment. The design of the pretreatment should be part of the process to feedwater quality adequacy for the EDR. The selection of the pre-treatment will depend on the wastewater composition and its variability, but will always comprise a step to remove solids and turbidity and another step to remove organic compounds. Usually physicochemical processes are used to remove solids and activated carbon filters to remove organics. As an alternative membrane processes and advanced oxidation may be used in the pretreatment. The dosage of chemicals during pre-treatment should be controlled to avoid residues that can enter in contact with EDR membranes.

During the process, fouling and scaling of membranes, electrodes and the entire system occurs. To minimize this problem, chemicals can be dosed, such as biocides, acids and anti-fouling agents, during operation or during the periodical cleaning of the system.

An analysis of chemical and operating conditions and the integrity of the installations, membranes, and electrodes, in addition to other preventive operational procedures, should be part of operational routine.

References

1. Allison RP (1995) Electrodialysis reversal in water reuse applications. Desalination 103(1–2):11–18. doi:10.1016/0011-9164(95)00082-8
2. Awwa Research Foundation (1996) Lyonnaise des Eaux-Dumez (Firm), South Africa Water Research Commission Electrodialysis. In: Water treatment membrane processes, McGraw-Hill, New York
3. Baker RW (2004) Ion exchange membrane processes—electrodialysis. Membrane technology and applications, 2nd edn. John Wiley & Sons, New York. doi: 10.1002/9781118359686.ch10
4. Broens L, Liebrand N, Futsellar H et al (2004) Effluent reuse at Barranco Seco (Spain): a 1,000 m3/h case study. Desalination 167:13–16. doi:10.1016/j.desal.2004.06.106
5. Cheremisinoff NP, Haddadin MB (2006) Beyond compliance—the refinery manager's guide to ISO 14001 implementation. Gulf Publishing Company, Houston, p 219
6. Elleuch M, Sistat P, Pourcelly G et al (2006) Brackish water desalination by electrodialysis: opposing scaling. Desalination 200(1–3):752–753. doi:10.1016/j.desal.2006.03.494
7. EPA—US Environmental Protection Agency (1992) Guideline for water reuse. types of reuse applications
8. EPA—US Environmental Protection Agency (2012) Guideline for water reuse. types of reuse applications
9. Gioli P, Silingardi GE, Ghiglio G (1987) High quality water from refinery waste. Desalination 67:271–282. doi:10.1016/0011-9164(87)90250-5
10. Lee HJ, Sarfert F, Strathmann H et al (2002) Designing of an electrodialysis desalination plant. Desalination 142(3):267–286. doi:10.1016/S0011-9164(02)00208-4
11. Leitz FB, Accomazzo MA, Mcrae WA (1974) High temperature electrodialysis. Desalination 14:33–41
12. Lozier CJ, Smith G, Chapman JW et al (1992) Selection, design, and procurement of a demineralization system for a surface water treatment plant. Desalination 88:3–31. doi:10.1016/0011-9164(92)80103-G
13. Machado BM (2008) Avaliação do Processo de Eletrodiálise Reversa no Tratamento de Efluentes de Refinaria de Petróleo. Máster Thesis, PPGE3 M, UFRGS, Brazil
14. Nicholas PC, Paul R (2009) The petroleum industry in handbook of pollution prevention and cleaner production. Elsevier Inc, Amsterdam, pp 1–96
15. Parkash S (2003) Refinery water system. In: Refining processes handbook, Elsevier, Amsterdam, pp 242–269. doi: 10.1016/B978-075067721-9/50009-8
16. Pilat B (2001) A case for electrodialysis. International water treatment, International Water & Irrigation
17. Reahl ER (2006) Half a century of desalination with electrodialysis. GE water and process technologies, Technical Paper. http://www.gewater.com. Accessed 10 Oct 2012

18. Reynolds TK, Leitz F (2004) Two years of operating experience at the port hueneme brackish water reclamation demonstration facility. In: AWWA, Water Desaltin Planing Guide for Water Utilities. John Wiley and Sons, Hoboken,. pp. 93–107
19. Rousseau RW (1987) Handbook of separation process technology. John Wiley & Sons. New York, p 1024
20. Ryabtsev AD, Kotsupalo NP, Titarenko VI et al (2001) Set-up involving electrodialysis for production of drinking-quality water from artesian waters with salt content up to 8 Kg/m^3 with productivity up to 1 m^3/h. Desalination 136:333–336. doi:10.1016/S0011-9164(01)00196-5
21. Ryabtsev AD, Kotsupalo NP, Titarenko VI et al (2001) Development of a two-stage electrodialysis set-up for economical desalination of sea-type artesian and surface waters. Desalination 137(1–3):207–214. doi:10.1016/S0011-9164(01)00220-X
22. Sato K, Kobayashi S, Okado S (1988) Desalination and reuse of industrial waste water by electrodialysis. Desalination 47(1–3):363–373. doi:10.1016/0011-9164(83)87092-1
23. Swami MSR, Muruganandam L, Mohan V (1996) Recycle of treated refinery effluents using electrodialysis-A case study. Indian J Environ Prot 16(4):282–285
24. Van der Hoek JP, Rijnbende DO, Lokin CJA et al (1998) Electrodialysis as an alternative for reverse osmosis in an integrated membrane system. Desalination 117(1–3):159–172. doi:10.1016/S0011-9164(98)00086-1
25. Walha K, Amar RB, Firdaous L, Quéméneur F, Jaouen P (2007) Brackish groundwater treatment by nanofiltration, reverse osmosis and electrodialysis in Tunisia: performance and cost comparison. Desalination 207(1–3):95–106. doi:10.1016/j.desal.2006.03.583
26. Watson IC, Morin OJ, Henthorne L (2003) Desalting handbook for planners, desalination and water purification, 3rd edn, Research and Development Program, Report n72. United States Department of the Interior. Bureau of Reclamation. Technical Service Center. Water Treatment Engineering and Research Group, p 310

Chapter 8
Electrodialysis Treatment of Tannery Wastewater

Kátia Fernanda Streit, Marco A. S. Rodrigues and Jane Zoppas Ferreira

Abstract The industrial processing of hides and skins consumes large volumes of water and generates waste that is highly polluted and causes environmental degradation. The conventional treatment of these effluents is not always effective in complying with the environmental regulations or in obtaining water with the necessary characteristics to be reused in the production process. In this sense, the search for new technologies to treat tanning effluents is undoubtedly a necessity and a great challenge for the industry. This chapter presents the application of electrodialysis (ED) as an alternative technique that can contribute to the reuse of water in the tanning production process, helping to minimize the environmental impact associated with the consumption of water and generation of effluents in the leather industry.

8.1 Introduction

The industrial processing of hides and skins involves four main stages [9]: (a) the beamhouse, which cleans and eliminates the substances that will not compose the leather; (b) the tanning, which is the stage where pretreated hides are transformed

K. F. Streit (✉)
Instituto Nacional de Metrologia, Qualidade e TecnologiaInmetro, Inmetro, Rio de Janeiro–RJ, Brazil
e-mail: kfstreit@yahoo.com.br

M. A. S. Rodrigues
Universidade FEEVALE, Novo Hamburgo–RS, Brazil
e-mail: MarcoR@feevale.br

J. Zoppas Ferreira
Programa de Pós Graduação em Engenharia de Minas, Metalúrgica e de Materiais (PPGE3M), Universidade Federal do Rio Grande do Sul (UFRGS), Porto Alegre–RS, Brazil
e-mail: jane.zoppas@ufrgs.br

A. Moura Bernardes et al. (eds.), *Electrodialysis and Water Reuse*, Topics in Mining, Metallurgy and Materials Engineering,
DOI: 10.1007/978-3-642-40249-4_8, © Springer-Verlag Berlin Heidelberg 2014

into leather; (c) the retanning, which complements the main tanning and confers some desired physical and mechanical properties to the leather, such as uniform color, tensile strength and softness, and (d) the finishing, which defines its final presentation and aspect. Each stage features several steps that include both chemical processes and mechanical operations and use large amounts of water and chemicals. The water used, along with the chemicals that do not react completely during the whole process, produces wastewater that is highly contaminated and responsible for environmental degradation [27]. These effluents are rich in organic matter, salts, tannins, chromium and nitrogen, among other substances.

Conventional treatment of these effluents is not always effective in complying with the environmental regulations or in obtaining water with the necessary characteristics to be reused in the production process.

Besides the legislation, which is becoming increasingly rigid with respect to environmental protection, the cost and the constant shortage of raw water has been a concern for the vast majority of hide and skin producers because of the important environmental impacts caused by water consumption and the generation of effluents in the leather industry.

Membrane separation processes, such as electrodialysis, have emerged as an alternative to conventional effluent treatment in tannery because it has the advantage of enabling the reuse of water and the chemicals present in the wastewater, reducing the demand for raw water and the amount of industrial waste generated.

8.2 Generation of Effluents in Tanneries

The world leather market process approximately 300 million hides per year [1]. The amount of water consumed and, consequently, the volume of wastewater generated in the processing of a salted bovine hide is approximately 1 m^3/hide [20]. This means that each year approximately 300 million m^3 of water are consumed and an equivalent volume of wastewater is generated in the worldwide production of leather. Considering an average daily per capita consumption of 150 $L{\cdot}day^{-1}$ in Brazil [6], the water consumed by tanneries represents an annual water intake of a population of 5.5 million inhabitants, which represents an important environmental impact of the leather industry.

8.3 Tannery Wastewater Characteristics

A review of the literature [3, 8, 11, 12, 22], IUE 6) shows that the inorganic contents of tannery wastewater are made up mainly of sodium, magnesium, chromium, chloride, ammonium, sulfate and sulfide. On the other hand, the organic fraction—usually composed by tannins and proteins, among others—is commonly expressed in terms of chemical oxygen demand (COD) and biochemical oxygen demand (BOD), or even in terms of total organic carbon (TOC).

The composition of tannery effluents is basically the following:

- Proteins derived from collagen, hair and interfibrillar proteins, besides those used in the finishing based on casein or albumin;
- The hide's own fat and oils used in greasing process;
- Non-ionic, cationic and anionic surfactants;
- Vegetable and synthetic tannins;
- Fungicides and bactericides based on phenol, halogenated phenols and triazole derivatives;
- Acrylic, polyurethane, urea and melamine resins;
- Acidic, basic and complex-metal dyes;
- Amines and enzymes;
- Chlorides, sulfates, phosphates, carbonates, bicarbonates, formiates, acetates, oxalates, citrates, lactates, sulfites and metabisulfites;
- Chromium, sodium, calcium, magnesium, aluminum, titanium and zirconium;
- Organic solvents, such as aliphatic hydrocarbons, glycol, isopropanol, ethanol, acetate and butyl acetate, methyl ethyl ketone and others.

It is important to mention that, after conventional treatment, some of these components are reduced to minimum levels. However, those considered most critical, such as tannins, nitrogen and some salts, remain in the treated effluent, damaging the waterways in which they are discharged and making it difficult to reuse the water in the production process.

Table 8.1 presents the characteristics of the industrial wastewater of tannery processes and of the final effluent. As can be seen, the beamhouse steps are responsible for the largest contribution in terms of inorganic and organic loads reaching the stream of the effluent treatment plant during the processing of hides and skins [3, 8, 11, 12, 22].

Effluents treated by a conventional process in 20 Brazilian tanning industries were collected and analyzed by an ISO 17025 accredited laboratory that specializes in the analysis of tannery wastewater. The parameters analyzed were Chemical Oxygen Demand (COD), conductivity, total nitrogen, and ammonium nitrogen. Conventional wastewater treatment is not very efficient in dealing with some parameters, especially nitrogen, and in this case the tannery industries are

Table 8.1 Tannery wastewater characteristics

Parameter	Beamhouse (kg/t)	Tanning (kg/t)	Retanning (kg/t)	Total (kg/t)	Final effluent (mg/L)
COD	135	5	20	160	5330
Chloride	135	30	2	167	5570
Sulfate	8	–	–	8	270
N-total	12	0.5	1	13.5	450
$N\text{-}NH_4^+$	3.5	0.2	0.8	4.5	150

Table 8.2 Tannery effluent characteristic after conventional treatment in Brazil [26]

Parameter	Concentration
DQO	100–400 $mgO_2 \cdot L^{-1}$
N-total	20–300 $mg \cdot L^{-1}$
$N\text{-}NH_4^+$	20–250 $mg \cdot L^{-1}$
Conductivity	1–12 $mS \cdot cm^{-1}$

facing difficulties to reach the required limits established by the regulations for Brazilian wastewaters published in 2005 [5]. The conductivity and COD parameters were analyzed in order to characterize the inorganic and organic fractions of the treated effluent. The range of values obtained is shown in Table 8.2.

8.4 Conventional Tannery Wastewater Treatment

The conventional treatment of tannery effluents consists of two steps: primary treatment, also called coagulation or physicochemical treatment, and the secondary treatment, also known as biological treatment [17]. Figure 8.1 presents a schematic of the treatment used by most tanning industries. In the primary treatment process, a coagulating agent is added to the wastewater. Aluminum salts such as $Al_2(SO_4)_3$ and $AlCl_3$ are commonly used. In solution, the aluminum ion hydrolyzes and generates some ionic species that interact with the wastewater contaminants, promoting the aggregation of colloidal particles and their sedimentation [7, 25]. The sediment fraction, or primary sludge, is then separated from the clear effluent and deposited in industrial landfills. This can cause the degradation of soil as well as of ground water.

The secondary step, or biological treatment, promotes oxidation of the organic contaminants. However, the capacity of some microorganisms to degrade some of the contaminants is limited, which means these compounds are not completely removed by the conventional biological process. For example, the tannins that are largely used in leather process have complex structures that are not easily degraded by microorganisms [16, 28]. Also, pH variations and effluent characteristics, such as type, structure and concentration of the contaminant, may inhibit microorganism metabolism [4]. Other limitations of the biological treatment include the long time needed for the process and the limited efficiency for removing color [18]. In addition, the presence of a broad spectrum of biocides, used in the leather industry to prevent fungal attack, may hinder the performance of microorganism [15].

Due to the limitations of conventional wastewater treatment in tannery, other processes have been pursued in the last years [2, 10, 11, 19, 24]. Amongst these, membrane separation processes such as electrodialysis have emerged as an alternative. This technique produces a diluted and a concentrated ion species

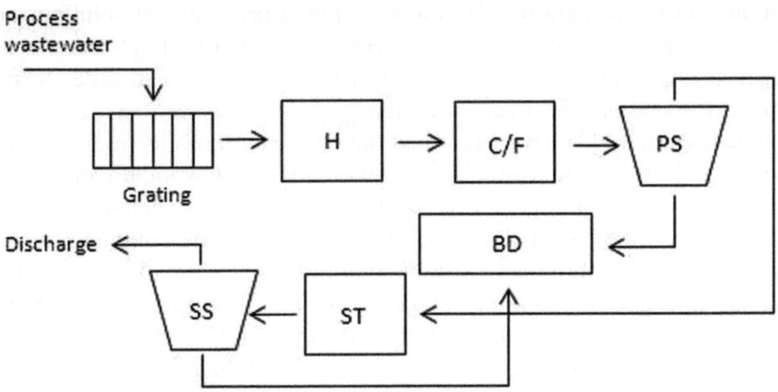

Fig. 8.1 Schematic of a conventional tannery effluent treatment plant. *H* homogeneizer; *C/F* coagulation/flocculation; *PS* primary sedimentation; *BD* bed drying; *ST* secondary treatment; SS: secondary sedimentation

stream, and has the advantage that, besides allowing the reuse of water, it may also enable the reuse of chemicals present in the effluent.

8.5 Electrodialysis to Process Wastewater

In the 1970s, the electrodialysis technique found one of its first applications in the leather industry. Mellon and colleagues studied the process [14] and used it for the separation of electrolytes in salted or pickled hides. Assisted by electrodialysis, they measured the acid and salt content of these hides.

Electrodialysis has been subsequently introduced in the tanning industry in order to recover water and chemicals for reuse in the production process. Especially in recent years, as a consequence of an increased demand for environmental care and also because of economic concerns.

Raghava Rao and colleagues [21] used electrodialysis to recover chromium salts and other neutral salts of residual tanning baths. In their study, the percentage of extraction observed was around 90 % for chloride and 50 % for sulfates, while the chromium content remained unchanged. This suggested that the described membrane separation procedure opened new possibilities for the reuse of water, neutral salts and chrome, without any problems in process control or effluent treatment.

The main limitation of the electrodialysis technique is fouling, which decreases the efficiency of the process over time.

Fouling is the accumulation of a deposit on the surface or within the membrane pores, caused by any species that can adhere on the membrane surface, reducing the capacity of ions diffusing through the membrane [13].

Fouling can be avoided by pH control, effluent pretreatment solutions and the addition of biocides, for example. However, even when fouling effects are minimized, it is important to define periodic cleaning processes in order to prolong membrane life and ensure process efficiency.

Studies in Brazil [23] have examined the application of electrodialysis to treat tannery effluents using the photoelectrooxidation (PEO) technique as pretreatment. The efficiency of the integrated PEO/ED system was evaluated by comparing parameters of the effluent before and after treatment. The combined techniques were able to remove more than 98 % of sodium, magnesium, chloride, sulfate and ammonium nitrogen. In addition, the performance of the PEO/ED treatment was also monitored in the wet blue leather retannage stage. Comparative tests were made by retanning shaved, wet blue bovine sided hides with a thickness of 1.2–1.4 mm for shoe uppers. Using the same recipe, one side was treated with normal feed water and the other side with cleared effluent water after PEO/ED treatment. The water obtained from this treatment presented similar characteristics to the feed water used in the tannery and its reuse in the process produced leather that was similar to those produced with the feed water, demonstrating the efficacy of electrodialysis in the treatment of tannery effluents. Table 8.3 presents the feed water and the effluent characteristics after PEO and electrodialysis (ED) treatment.

In another study conducted in Brazil and Portugal [26, 27], the application of electrodialysis to treat tannery effluents using the nanofiltration (NF) technique as pretreatment was evaluated. This work was conducted with model solutions that simulated the tannery effluent after conventional treatment. A membrane cleaning process using NaOH and Na_2CO_3 was also studied and the experiments showed that a cleaning process every 15 days is enough to ensure the performance of electrodialysis.

Table 8.4 presents the contents of the model solution used before and after the application of the electrodialysis technique with new, used and cleaned membranes. Although the electrodialysis technique presented lower efficiency for the

Table 8.3 Feed water and effluent characteristics [23]

Parameter	Feed water	Effluent after PEO/ED treatment
COD (mgO_2L^{-1})	8.0	19.0
BOD (mgO_2L^{-1})	6.0	4.0
pH	7.8	5.5
Total solids (mgL^{-1})	142	128
Total nitrogen (mgL^{-1})	7.8	4.8
Ammonium nitrogen (mgL^{-1})	5.4	<0.2
Chloride (mgL^{-1})	10.4	9.3
Phosphorus (mgL^{-1})	0.04	0.05
Total cromium (mgL^{-1})	<0.01	<0.01
Total cálcium (mgL^{-1})	9.8	0.3
Total magnesium (mgL^{-1})	5.4	0.08
Total sodium (mgL^{-1})	8.8	26.5

Table 8.4 Concentration of the model solutions [27]

Parameter	Model Solution	After ED with new membranes	After ED with used membranes	After ED with cleaned membranes
COD ($mgO_2 \cdot L^{-1}$)	312 ± 10	10.0 ± 1.0	10.9 ± 1.0	12.0 ± 1.0
Conductivity ($mS \cdot cm^{-1}$)	4.67 ± 0.15	3.28 ± 0.15	3.51 ± 0.17	3.67 ± 0.14
Ammonium nitrogen ($mg \cdot L^{-1}$)	88.0 ± 2.0	58.1 ± 2.0	59.0 ± 2.5	57.2 ± 1.5
Sodium ($mg \cdot L^{-1}$)	458 ± 15	330 ± 13	349 ± 13	344 ± 8
Magnesium ($mg \cdot L^{-1}$)	365 ± 15	182 ± 16	256 ± 16	219 ± 11
Chloride ($mg \cdot L^{-1}$)	840 ± 25	386 ± 24	487 ± 19	437 ± 15
Suphate ($mg \cdot L^{-1}$)	770 ± 25	578 ± 22	655 ± 22	647 ± 22

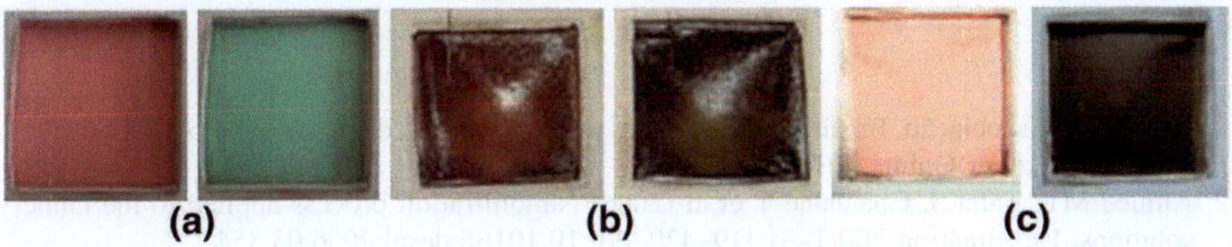

Fig. 8.2 Membranes used in the ED experiments: **a** new membranes, **b** used membranes and **c** cleaned membranes after 15 days

membranes used for 15 days, the cleaned membranes presented results that were similar to the new ones in ED experiments.

Figure 8.2 shows the cationic (left) and anionic (right) membrane pairs used in the ED experiments. It can be observed that, after cleaning (c), the cationic membrane has a similar appearance to the new membrane (a). When the anionic membrane is considered, one can see that the cleaning process is not as efficient from a visual perspective, because this (anionic) membrane is more prone to fouling [13].

8.6 Concluding Remarks

The electrodialysis technique enabled the removal of COD, conductivity, sodium, magnesium, chloride and sulfate and the residual levels of these components in the treated effluent indicate the possibility of complying with the standards set by legislation and the requirements for reuse in the tanning industry, including in its most critical steps, such as retanning, dyeing and greasing, where water characteristics are key to the quality of the leather produced.

Organic matter contained in the effluent can decrease the performance of the system, causing the fouling of membranes and decreasing the percentage of extraction of ions. Pretreatment may therefore be important to improve the results. The alkaline cleaning process has proven to be effective in minimizing potential problems related to fouling and in increasing the useful life of membranes. This avoids unnecessary costs associated with the frequent replacement of the membranes.

In view of this, the use of membrane separation processes, such as electrodialysis, in the treatment of tannery effluents is technically feasible when the objective is improving the characteristics of the conventionally treated effluent. Electrodialysis enables the reuse of effluents in the production process, and helps to minimize the environmental impact caused by the water consumption and effluent discharge associated with the leather industry.

References

1. ABQTIC- Associação Brasileira dos Químicos e Técnicos da Indústria do Couro (2012) Brazilian Leather Guide. ABQTIC, Brasil
2. Ahmed MT, Taha S, Chaabane T et al (2006) Nanofiltration process applied to the tannery solutions. Desalination 200(1–3):419–420. doi:10.1016/j.desal.2006.03.354
3. Aleixandre MVG, Roca JAM, Bes-Piá A (2011) Reducing sulfates concentration in the tannery effluent by applying pollutionprevention techniques and nanofiltration. J Clean Prod 19:91–98. doi:10.1016/j.jclepro.2010.09.006
4. Bertazzoli R, Pelegrini R (2002) Photoelectrochemical discoloration and degradation of organic pollutants in aqueous solutions. Quim Nova 25:477–482
5. Brazil (2005) Ministério do Meio Ambiente, Conselho Nacional do Meio Ambiente, CONAMA, Resolução 357, de 17 de março de 2005, Dispõe sobre a classificacão dos corpos de água e diretrizes ambientais para o seu enquadramento, bem como estabelece as condições e padrões de lançamento de efluentes, e dá outras providências
6. Brazil (2010) Sistema Nacional de Informações sobre Saneamento. Diagnóstico dos serviços de água e esgoto. htpp://www.snis.gov.br. Accessed 01 Apr 2013
7. Cañizares P, Martínez F, Jiménez C et al (2006) Comparison of the aluminum speciation in chemical and electrochemical dosing processes. Ind Eng Chem Res 45:8749–8756
8. Cassano A, Adzet J, Molinari R et al (2003) Membrane treatment by nanofiltration of exhausted vegetable tannin liquors from the leather industry. Water Res 37(10):2426–2434. doi:10.1016/S0043-1354(03)00016-2
9. Costa CR, Botta CMR, Espindola ELG et al (2008) Electrochemical treatment of tannery wastewater using DSA® electrodes. J Hazard Mater 153(1–2):616–627. doi:10.1016/j.jhazmat.2007.09.005
10. Das C, De S (2009) Steady state modeling for membrane separation of pretreated liming effluent under cross-flow mode. J Membr Sci 338(1–2):175–181. doi:10.1016/j.memsci.2009.04.029
11. Gisi S, Galasso M, De Feo G (2009) Treatment of tannery wastewater through the combination of a conventional activated sludge process and reverse osmosis with a plane membrane. Desalination 249(1):337–342. doi:10.1016/j.desal.2009.03.014
12. IUE–International Union of Environment (2008) International Union of Leather Technologists in Chemists Societies-IULTCS. Document on typical pollution values related to conventional tannery processes–IUE 6

13. Lindstrand V, Sundstrom G, Jonson A-S (2000) Fouling of electrodialysis membranes by organic substances. Desalination 128(1):91–102. doi:10.1016/S0011-9164(00)00026-6
14. Mellon FE, Gruber AH, Staller VM (1972) An electrodialysis cell for the determination of total electrolyte in cured or pickled hides. J.A.L.C.A. 67:5–15
15. Meneses ES, Arguelho MLPM, Alves JPH (2005) Electroreduction of the antifouling agent TCMTB and its electroanalytical determination in tannery wastewaters. Talanta 67(4):682–685. doi:10.1016/j.talanta.2005.01.058
16. Mingshu L, Kai Y, Qiang H et al (2006) Biodegradation of gallotannins and ellagitannins. J Basic Microbiol 46:68–84
17. Munz G, De Angelis D, Gori R et al (2009) The role of tannins in conventional and membrane treatment of tannery wastewater. J Hazard Mater 164(2–3):733–739. doi:10.1016/j.jhazmat.2008.08.070
18. Nogueira RFP, Jardim WF (1998) Heterogeneous photocatalysis and its environmental applications. Quim Nova 21:69–72
19. Prabhavathy C, De S (2010) Treatment of fatliquoring effluent from a tannery using membrane separation process: Experimental and modeling. J Hazard Mater 176(1–3):434–443. doi:10.1016/j.jhazmat.2009.11.048
20. Rao JR, Chandrababu NK, Muralidharan C et al (2003) Recouping the wastewater: a way forward for cleaner leather Processing. J Clean Prod 11(5):591–599. doi:10.1016/S0959-6526(02)00095-1
21. Rao JR, Prasad BGS, Narasimhan V et al (1989) Electrodialysis in the recovery and reuse of chromium from industrial effluents. J Membr Sci 46(1–3):215–224. doi:10.1016/S0376-7388(00)80336-1
22. Renner G. Abwässer aus Gerbereien und Möglichkeiten der Belastungsverminderung (2004) Veränderung der organischen Belastung von Gerbereiabwasser durch biologische Behandlung. In: Abwässer aus der Zellstoffindustrie und der Lederherstellung. Colloquium TU Berlin, pp 187–201
23. Rodrigues MAS, Amado FDR, Xavier JLN et al (2008) Application of photoelectrochemical-electrodialysis treatment for the recovery and reuse of water from tannery effluents. Journal of Cleaner Production 16:605–611. doi:org/10.1016/j.jclepro.2007.02.002
24. Sahu SK, Meshram P, Pandey BD et al (2009) Removal of chromium(III) by cation exchange resin, Indion 790 for tannery waste treatment. Hydrometall 99(3–4):17–170. doi:10.1016/j.hydromet.2009.08.002
25. Song Z, Williams CJ, Edyvean RGJ (2004) Treatment of tannery wastewater by chemical coagulation. Desalination 164:249–259. doi:10.1016/S0011-9164(04)00193-6
26. Streit KF, Ferreira JZ, Bernardes et al (2009) Ultrafiltration/nanofiltration for the tertiary treatment of leather industry effluents. Environ Sci Technol 43:9130–9135. doi: 10.1021/es902105q
27. Streit KF, Gerevini G, Rodrigues MAS et al (2013) Electrodialysis in an Integrated NF/ED Process for Water Recovery in the Leather Industry. Sep Sci Technol 48:445–454
28. Szpyrkowicz L, Kaul SN, Neti RN et al (2005) Influence of anode material on electrochemical oxidation for the treatment of tannery wastewater. Water Res 39(8):1601–1613. doi:10.1016/j.watres.2005.01.016

Chapter 9
Electrodialysis Treatment of Phosphate Solutions

Daniel Arsand and Andréa Moura Bernardes

Abstract An effective way to prevent corrosion of metals and alloys is to coat the substrate material. The phosphating process is usually used to provide a phosphate layer for subsequent application of organic coating systems to metal surfaces, such as steel, zinc, aluminum and others. The phosphating process includes several rinsing processes and ions such as PO_4^{3-}, Fe^{2+}, Ni^{2+} and Zn^{2+} are found in the generated wastewater. Emission reduction and eutrophication control require wastewater treatment to remove phosphorous. There is a range of methods available for reducing phosphate in wastewater. Phosphating effluents are usually treated by physicochemical methods, but biological operations are also frequently used. The concentrations of metallic ions are normally above the environmental limits and the coagulation/precipitation technique is usually applied for their removal. However, the use of membranes in tertiary treatment is a promising technology. The rinse water of the phosphating process contains high concentrations of salts and low concentrations of organic matter, which means that electrodialysis treatment may be an attractive solution.

9.1 Introduction

Phosphorus is required by every living plant and animal cell. However, it is not present in abundance on the earth's surface. It occurs almost exclusively in nature as orthophosphate (PO_4^{3-}), making it suitable for biological metabolism.

D. Arsand (✉)
Instituto Federal Sul Riograndense, Pelotas–RS, Brazil
e-mail: Daniel.arsand@gmail.com

A. Moura Bernardes
Programa de Pós Graduação em Engenharia de Minas, Metalúrgica e de Materiais (PPGE3M), Universidade Federal do Rio Grande do Sul (UFRGS), Porto Alegre–RS, Brazil
e-mail: amb@ufrgs.br

A. Moura Bernardes et al. (eds.), *Electrodialysis and Water Reuse*,
Topics in Mining, Metallurgy and Materials Engineering,
DOI: 10.1007/978-3-642-40249-4_9,

Orthophosphate can be present as PO_4^{3-}, HPO_4^{2-}, $H_2PO_4^-$ and H_3PO_4, depending on the pH [1]. The phosphate is a key fertilizer ingredient in agricultural production, but it is also a known pollutant. Excess phosphorus in surface waters can stimulate the growth of algae, which in turn reduces the oxygen concentration in water bodies, leading to adverse environmental consequences. Phosphate is recognized as one of the major nutrients that contribute to the worldwide increase in eutrophication [2]. In addition to the use of phosphate as agricultural fertilizer, this compound can be found in domestic and industrial wastewater. Industry, where phosphating operations are commonly applied in the preparation of metals in metal painting, is an important source of phosphorus production [3]. In the metal-mechanics industry, especially in the plating and phosphating sector, large amounts of wastewater are generated containing chemical additives, nutrients and heavy metals. In order to reduce emissions and control eutrophication, wastewater treatment must be able to remove phosphorous and heavy metals. As a consequence, wastewater treatment plants are necessaries.

Traditionally, biological and physicochemical technologies are employed for the removal of phosphates. However, new technologies are emerging that are appropriate and which produce better results [2, 4] Membrane technologies, for example, are an alternative method of wastewater treatment and can be applied to treat phosphating wastewater [5].

9.2 The use of Phosphate Solutions in the Metal-Mechanics Industry

Acid phosphoric and phosphate solutions are frequently used in the metal-mechanics industries. The phosphating treatment provides a base for organic coating procedures of metal surfaces and the phosphate layer and painting combination provides better protection against corrosion.

The phosphating process is associated to an aqueous phosphoric acid solution (pH 2.5–3.2), which, in addition to phosphates, usually contains nitrate and other ions, such as zinc, manganese, calcium, nickel, and iron [6–8].

The phosphating process requires mainly iron products, but zinc and aluminum are also used. The process consists of immersing or a spray treating the metal substrate with the aqueous solution [9].

The phosphating process includes degreasing, pickling and phosphating steps. After each step, the metal product is rinsed with water. The phosphating process, therefore, includes several rinsing steps, which generates wastewater with phosphoric acid, zinc, nickel and manganese as components [10, 11].

9.3 Conventional Phosphate Wastewater Treatment

Industrial programs have focused on end-of-pipe solutions to control the discharge of wastewater pollutants. Phosphating wastewater treatment plants are no different. The main objective is to remove metallic and phosphate ions. There are a range of methods available for reducing phosphate concentrations in wastewater. Chemical removal with iron or aluminum salts and pH adjustment is a reliable method, which causes the precipitation of phosphorus and metal hydroxides, producing a sludge.

The pH used for the chemical coagulation to remove metals in effluents is close to the pH used to precipitate phosphate ions (Eq. 9.1). Precipitation of phosphate salts is an important step in regulating the level of phosphorous in natural water systems. Hydroxyapatite, a substance with lower solubility, is formed when a pH equal to 10 is used to precipitate phosphate ions (Eq. 9.2).

$$H_3PO_4 + Ca(OH)_2 \rightarrow CaHPO_4 + 2H_2O \tag{9.1}$$

$$3H_3PO_4 + 5Ca(OH)_2 \rightarrow Ca_5(PO_4)_3(OH) + 9H_2O \tag{9.2}$$

However, after treating the effluents, residual phosphate ions are still present (HPO_4^- and PO_4^{3-}). Usually, this means that the treated wastewater requires a tertiary treatment to remove the phosphate (Fig. 9.1).

Filtration and/or flotation are also used. It is possible to combine coagulation agents with filtration/flotation for a more efficient removal of phosphate. Such techniques as electrocoagulation can be used. Kobya et al. [4] obtained 98 % phosphate and zinc removal from phosphating wastewater by electrocoagulation.

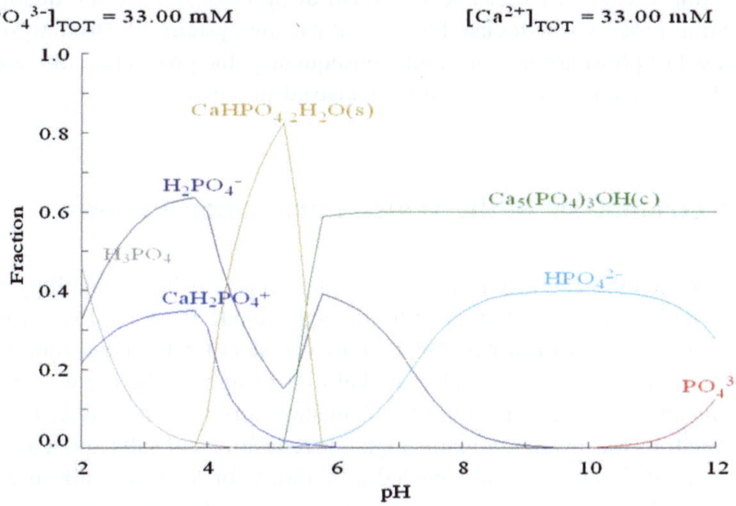

Fig. 9.1 Chemical Equilibrium Diagram of Phosphate and Calcium Ions

Liu et al. [12] studied the influence of pH and sulphate concentrations on phosphate removal with calcite. The removal with calcite was favorable in acidic and strong basic solutions. The presence of sulfate significantly influenced phosphate removal with calcite. In addition, low concentrations of sulfate in acidic conditions can promote phosphate removal, but high sulfate concentrations can inhibit phosphate removal because calcium sulfate is formed first [12].

Currently, iron phosphate produced through the chemical phosphate removal process cannot be used to replace mined phosphate for most industrial uses. Furthermore, iron phosphate is likely to have relatively low bioavailability under aerobic conditions (such as in soil) for plants. This reduces its suitability for replacing phosphate fertilizers [13].

Biological processes to remove phosphate are also applied. When biological phosphate removal is used, an additional step for the recovery of such phosphates as struvite (a phosphorus mineral) can be added. Struvite can be a source of phosphorus, nitrogen and magnesium for plants. The recovery and use of the phosphorus in struvite also offers a number of major sustainable advantages, including the protection of natural resources of phosphate rocks and environmental enhancements through nutrient recovery/recycling. In areas with significant variability of process parameters (e.g. the flow rate and composition changes due to heavy rainfalls), chemical methods for the removal of phosphorus might offer further advantages in comparison to biological methods [13]. Regardless of the method used, however, the chemical and biological treatment of phosphating wastewaters will produce a sludge that also contains metals, which makes its reutilization as a nutrient more difficult.

Taking into account that phosphating wastewater is usually treated by physicochemical and biological methods, and that the treated wastewater is not good enough to be reused as industrial water, appropriate technologies must be applied to make this wastewater reusable in industrial processes, since the quality of the phosphating process is associated to a good rinsing system. Electrodialysis allows the removal of phosphorous salts and, consequently, the production of water that is compatible in quality for the reuse in industrial processes.

9.4 Electrodialysis in the Water Purification Process

The use of membranes in tertiary treatment is a promising technology. Reverse osmosis, nano-, ultra- and micro-filtrations have all been explored to remove contaminants from wastewater and to purify water. Chilyumova and ThÖming [14] studied the retention of bivalent nickel ions from phosphating wastewater and the results showed a higher retention of bivalent ions than monovalent ones.

Phosphating wastewater contains high concentrations of salts and a low organic matter content. These are the expected conditions for a successfully use of electrodialysis, which has become increasingly attractive for treatment and recycling of wastewater in industries that use the phosphating process.

The cations that are present in higher concentrations in phosphating wastewater are Fe^{2+}, Ni^{2+} and Zn^{2+}. To reuse the wastewater as rinse water, these metallic ions must be removed. Different studies have been carried out to remove metal by electrodialysis. Results obtained by [15] demonstrated a recovery of 90 % of Ni^{2+} and Zn^{2+} using electrodialysis. The treatment of an acid solution can reach 82–91 % Fe^{2+} removal with 76–80 % of current efficiency [16]. A performance with about 97 % removal of heavy metals, 95 % of anions and 85 % of COD (Chemical Oxygen Demand), with final heavy metal concentrations below 0.01 mg L^{-1}, was obtained by [5].

Other authors have studied the recovery of phosphates by electrodialysis. (Kim et al. [17]) applied electrodialysis to obtain solutions with NH_4^+-N and PO_4^{3-}-P concentrations of about 96 and 94 %, respectively. Phosphoric acid (89.2 % purity) was recovered from wastewater containing phosphate salts by employing bipolar membranes with a current efficiency of 95.2 % [18].

The wastewater composition can modify the separation efficiency. The separation efficiency changed after 2 h of industrial effluent treatment because of the phosphate, sulfate, nitrate, chloride and other anions present in feed effluent [15]. The treatment time changes according to the composition of the medium [19]. The effluent composition, and consequently the ion combinations, influences the separation performance in electrodialysis, since the ligand exchange reactions affect the transport properties of ions [20].

The extraction of iron II, nickel II and zinc II ions has been studied in synthetic phosphate effluents. The studies were conducted with solutions containing phosphates and only one type of metal ion, as well as with solutions containing all three types of ions. Under the same experimental conditions, the percentage extraction for iron ions treated in the absence of other ions was lower than the extraction obtained when iron was treated with nickel and zinc ions. In contrast, nickel ions in the working solution containing only other nickel ions were extracted more efficiently than when nickel was treated with iron and zinc. This is probably associated to an ion competition [21]. In addition, complexes are formed between iron and phosphate ions, which can influence the transport characteristics (Fig. 9.2). Studies by Rodrigues et al. [22] have shown a relationship between the ions in solution and the formation of different species, modifying the ion extraction on the treatment by electrodialysis.

Electrodialysis may be combined with other techniques to reach better results. Badruzzaman et al. [23] showed an alternative use of electrodialysis, combining it with reverse osmosis to generate acids, bases and hypochlorite solutions. Multiple membrane separations in the treatment of plating wastewater have been introduced in the production line. Electrodialysis has been also coupled on-line in the platting process. An integrated microfiltration-electrodialysis system has been proposed for wastewater reclamation containing phosphate salts. The concentration ratio of the concentrate and dilutant will determine if a better separation is achieved. In order to obtain higher separation efficiency and cut operating costs, the concentrate-dilutant ratio should be controlled around 7 [5].

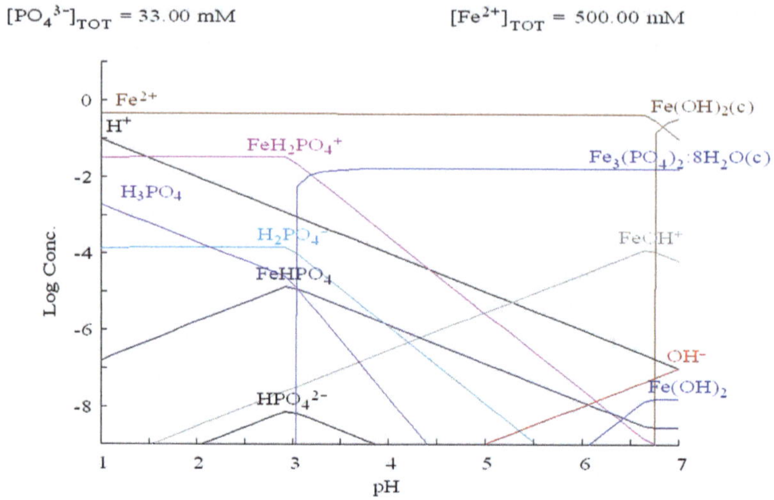

Fig. 9.2 Chemical Equilibrium Diagram of Phosphate and Iron Ions

9.5 Fouling and Scaling in Phosphating Wastewater Treatment

Common problems that must be controlled during electrodialysis treatment are fouling and scaling. When fouling or scaling occurs, the process must be stopped and mechanical cleaning and treatment with dilute bases and acids are necessary to restore the original properties of the membranes.

Phosphating wastewater may contain organic matter and promote fouling. Hydrophobic interactions between organic molecules and the membrane polymer can result in a strong and irreversible surface adsorption. The absence of organic matter is therefore very important in phosphating wastewater treatment. The presence of organic matter in the feed water of electrodialysis decreases the efficiency of the process because fouling and biofouling of the exchange membrane can occur, increasing the resistivity of the system. A few methods can be used to inhibit fouling in phosphating wastewater treatment using electrodialysis. The use of ozonation combined with electrodialysis has been shown to reduce fouling [17]. Ultrafiltration can be applied to separate the high molecular weight fraction. The kind of membrane used affects fouling. The anion-exchange *membrane* is more sensitive to *fouling*. Ion-exchange membranes based on aliphatic polymers are less sensitive to fouling than membranes based on aromatic polymers [24].

Precipitation of salts on the membrane surface, on the other hand, results in an increase of the electrical resistance of the stack and may eventually lead to physical damage of the membrane. The limiting current density is therefore a significant parameter that must be determined experimentally [25] to work in optimized conditions.

The occurrence of scaling in phosphating wastewater treatment is possible and controlling the current density is especially important in the treatment of wastewater containing iron. The dissociation of water leads to the precipitation of metallic hydroxides on the cation exchange-membranes [26]. The applied current must be lower than the limit current density for a more efficient removal of iron. The increasing applied current results in higher ions removal, however current efficiency decay can be observed. The low concentration of ions is one reason for this behavior. Normally, when a low current efficiency is observed, the temperature increases. Electrosmose in these cases are commonly observed. Percentages up to 25 % were noted when treating synthetic phosphating rinse water.

Although scaling is usually a consequence of the precipitation of metallic hydroxides, anions in the wastewater can also precipitate. Insoluble calcium phosphate salts precipitate on the concentrating surface of the anion exchange membrane. However, scaling may be avoided with applied current below that limiting current density or with chemical products additions. To inhibit salt precipitation on the membrane surface, a controlling agent can be used. Calcium carbonate and calcium sulphate precipitation on the membrane surface can be controlled by $Na_2[Na_4(PO_3)_6]$ (Eqs. 9.3 and 9.4).

$$Na_2\left[Na_4(PO_3)_6\right] + CaCO_3 \leftrightarrow Na_2\left[Na_2Ca(PO_3)_6\right] + Na_2CO_3 \tag{9.3}$$

$$Na_2\left[Na_4(PO_3)_6\right] + CaSO_4 \leftrightarrow Na_2\left[Na_2Ca(PO_3)_6\right] + Na_2SO_4 \tag{9.4}$$

9.6 Concluding Remarks

The phosphating process includes several rinsing processes and ions such as Fe^{2+}, Ni^{2+}, Zn^{2+} and PO_4^{3-} are found in the generated wastewater. Emission reduction and eutrophication control require wastewater treatment to be able to remove heavy metals and phosphorous.

Electrodialysis is employed in the treatment of industrial wastewater and can be applied to the treatment of phosphating wastewater. This technique shows advantages over other membrane processes. Electrodialysis presents high membrane stability, high ionic separation and low energy consumption, important parameters in the recovery of wastewater. The treatment of phosphating wastewater by combining ED and other processes, such as reverse osmosis, filtration technologies and ozonation, can be used to reach better results.

References

1. Chapra SC (1997) Surface water quality modeling. WCB/McGraw-Hill. 844p
2. Golder AK, Samanta AN, Ray S (2006) Removal of phosphate from aqueous solutions using calcined metal hydroxides sludge waste generated from electrocoagulation. Sep Purif Technol 52(1):102–109. doi:10.1016/j.seppur.2006.03.027
3. McComas C, McKinley D (2008) Reduction of phosphorus and other pollutants from industrial dischargers using pollution prevention. J Clean Prod 16(6):727–733. doi:10.1016/j.jclepro.2007.02.023
4. Kobya M, Demirbas E, Dedeli A et al (2010) Treatment of rinse water from zinc phosphate coating by batch and continuous electrocoagulation processes. J Hazard Mater 173(1–3):326–334. doi:10.1016/j.jhazmat.2009.08.092
5. Zuo W, Zhang G, Meng Q (2008) Characteristics and application of multiple membrane process in plating wastewater reutilization. Desalination 222(1–3):187–196. doi:10.1016/j.desal.2007.01.149
6. Zucchi F, Trabanelli G (1971) Anodic Behavior of Fe in phosphate solutions. Corros Sci 11:141–151. doi:10.1016/S0010-938X(71)80090-2
7. Burokas V, Martušienė A, Girčienė O (2007) Influence of fluoride ions on the amorphous phosphating of aluminium alloys. Surfac Coat Technol 202(2):239–245. doi:10.1016/j.surfcoat.2007.05.034
8. Zhang S-I, Chen H–H, Zhang X-L et al (2008) The growth of zinc phosphate coatings on 6061-Al alloy. Surf Coat Technol 202(9):1674–1680. doi:10.1016/j.surfcoat.2007.07.037
9. Li GY, Lian JS, Niu LY et al (2006) Growth of zinc phosphate coatings on AZ91D magnesium alloy. Surf Coat Technol 201:1814–1820. doi:10.1016/j.surfcoat.2006.03.006
10. Juchi K, Ying H (2010) Effect of Rare Earth on the coating-forming and mechanism of phosphatization. J Rare Earths 28:132. doi:10.1016/S1002-0721(10)60352-3
11. Juchi K (2007) Study of melioration of stainless steel surface by phosphatisation with RE additives and mechanism. J Rare Earth 25:275. doi:1002-0721(2007)-0275-06
12. Liu Y, Sheng X, Dong Y et al (2012) Removal of high-concentration phosphate by calcite: Effect of sulfate and pH. Desalination 289:66–71. doi:10.1016/j.desal.2012.01.011
13. Fuentes B, de la Luz Mora M, Bolan NS et al (2008) Chapter 16—Assessment of phosphorus bioavailability from organic wastes in soil. Dev Soil Sci 32:363–411
14. Chilyumova E, Thöming J (2008) Nanofiltration of bivalent nickel cations—model parameter determination and process simulation. Desalination 224:12–17. doi:10.1016/j.desal.2007.04.072
15. Santarosa VE, Peretti F, Caldart V et al (2002) Study of ion-selective membranes from electrodialysis removal of industrial effluent metals II: Zn and Ni. Desalination 149:389–391
16. Wisniewska G, Wiśniewski J, Winnicki T (1993) Recovery of acid and water by Membrane Dialysis. Desalinations 91:307–317
17. Kim JO, Jung JT, Chung J (2007) Treatment performance of metal membrane microfiltration and electrodialysis integrated system for wastewater reclamation. Desalination 202:343–350. doi:10.1016/j.desal.2005.12.073
18. Trivedi GS, Shah BG, Adhikary SK et al (1999) Studies on bipolar membranes Part III: conversion of sodium phosphate to phosphoric acid and sodium hydroxide. React & Funct Polym 39(1):91–97. doi:10.1016/S1381-5148(97)00159-4
19. Kabay N, İpek Ö, Yüksel M (2006) Effect of salt combination on separation of monovalent and divalent salts by electrodialysis. Desalination 198:84–91. doi:10.1016/j.desal.2006.09.013
20. Rodrigues MAS, Dalla Costa RF, Bernardes AM et al (2001) Influence of ligand exchange on the treatment of trivalent chromium solutions by electrodialysis. Electrochim Acta 47(5):753–758. doi:10.1016/S0013-4686(01)00756-3
21. Arsand D (2001) Tratamento de Efluentes de Fosfatização por Eletrodiálise. Máster Thesis,PPGE3 M, UFRGS, Brazil

22. Rodrigues MAS, Amado FDR, Bischoff MR et al (2008) Transport of zinc complexes through ananion exchange membrane. Desalination 227(1–3):241–252
23. Badruzzaman M, Oppenheimer J, Adham S et al (2009) Innovative beneficial reuse of reverse osmosis concentrate using bipolar membrane electrodialysis and electrochlorination processes. J Membr Sci 326:392–399. doi:10.1016/j.memsci.2008.10.018
24. Noble RD, Stern SA (1995) Membrane Separations Technology—Principles and Applications. Elsevier, Amsterdan
25. Lee H-J, Sarfert F, Strathmann H et al (2002) Designing of an electrodialysis desalination plant. Desalination 142:267–286
26. Tanaka Y (2007) Acceleration of water dissociation generated in an ion exchange membrane. J Membr Sci 303:234–243. doi:10.1016/j.memsci.2007.07.020

Chapter 10
Electrodialysis for the Recovery of Hexavalent Chromium Solutions

Christa Korzenowski, Marco A. S. Rodrigues and Jane Zoppas Ferreira

Abstract The widespread use of chromium and its compounds in modern industry results in the discharge of great quantities of this element into the environment. Conventional waste treatment technologies are usually based on the transfer of the contamination to another waste product with the so called "end of pipe technologies". The storage of hazardous solid waste in dump sites presents a serious health and safety risk to the community, and other disposal methods are required. It seems prudent to consider waste as a resource that can be detoxified and converted to usable products. This chapter provides a short review of the hexavalent chromium electroplating process, showing the characteristics of the process and the contaminants generated by it. It also shows the types of treatment that are most commonly used by industries and the new treatment technologies that are considered to be cleaner, such as electrodialysis.

10.1 Introduction

Today, chromium is one of the strategic metals and it is, crucial for the military survival and economic welfare of all industrialized nations. It is now one of the most important elements in the production of modern alloys and plays a key role in all major technological developments. It is widely used as an alloy with iron to

C. Korzenowski (✉) · J. Zoppas Ferreira
Programa de Pós-Graduação em Engenharia de Minas, Metalúrgica e de Materiais (PPGE3M), Universidade Federal do Rio Grande do Sul (UFRGS), Porto Alegre–RS, Brasil
e-mail: ckorzenowski@gmail.com

J. Zoppas Ferreira
e-mail: jane.zoppas@ufrgs.br

M. A. S. Rodrigues
Universidades Feevale, Novo Hamburgo–RS, Brasil
e-mail: marcor@feevale.br

A. Moura Bernardes et al. (eds.), *Electrodialysis and Water Reuse*,
Topics in Mining, Metallurgy and Materials Engineering,
DOI: 10.1007/978-3-642-40249-4_10,

give steel the combined properties of high hardness, toughness and great resistance to chemical attack, and it is a major constituent of stainless steel. Chromium in metallic form is extremely resistant to corrosion and high hardness is one of its common features. It's widely employed as a protective layer electroplated over other metals. Hexavalent chromium is an extremely polluting and carcinogenic metal, and its use is even banned in some countries for some applications. However, its use in the chromium electroplating processes is still permitted and, therefore, it is still widely used. This requires large investments in wastewater treatment and measures to prevent accidents. Liquid effluents of electroplating industries that use chromium processes consist basically of water from the rinsing steps and the disposal of exhausted plating baths, generating a highly toxic effluent. The disposal of these effluents generates large amounts of chromium in the environment, and therefore deserves special attention, mainly because they contain chromium in the hexavalent form [1, 2].

10.2 The Use of Hexavalent Chromium in the Electroplating Processes

Chrome plating is a finishing treatment usually applied to ferrous metals. The chromium can be electroplated onto surfaces for decorative effects, or to improve some surface properties (hardness and corrosion resistance). It is resistant to heat and is not subject to fogging. The thickness of electroplated chromium deposits falls into two classifications: decorative and functional. The decorative chromium plating process consists in depositing a thin layer of chromium, in the order of 1–2 μm, usually applied to pre-deposits of nickel or copper. They offer a pleasing, reflective appearance, while also providing corrosion resistance, lubricity and durability. Hard chromium plating, which seeks to improve the mechanical properties, is obtained by applying thicker layers, generally greater than 20 μm, and is used in industrial, non-decorative applications. In contrast to decorative deposits, functional chromium is usually plated directly on the substrate and only occasionally over electrodeposits, such as nickel or copper [3].

Figure 10.1 presents a flow chart of a common decorative chromium plating process. It is possible to see that the main liquid effluents are generated during the rinsing steps associated with the plating process.

10.3 Characteristics and Contaminants of Hexavalent Chromium Solutions

Electrochemical processes in systems with chromic oxide (CrO_3) dissolved in water as main component, are widely used in electroplating. Sulfate ion (SO_4^{2-}) is a necessary catalyst in all chromium plating solutions. It is usually introduced as

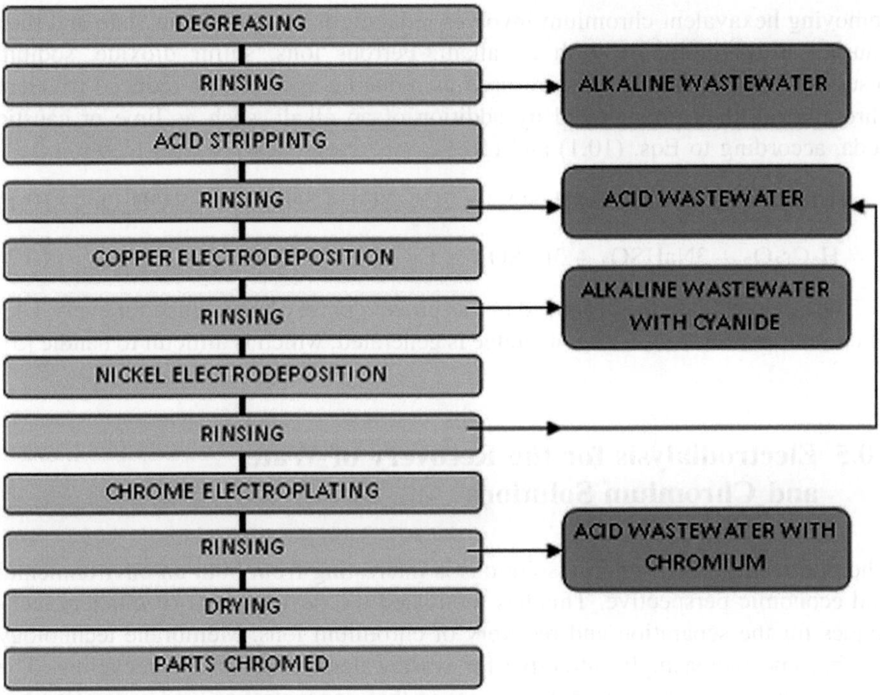

Fig. 10.1 Decorative chromium plating process flow chart

sulfuric acid. In the chromium plating industry, the solutions contain chromate concentration varying between 250 and 450 gL^{-1}, depending on which bath is used. Less than 10 % of the chromium acid used is deposited on the metal products. After a period of use, these chromium acid solutions become contaminated with metals such as Fe^{3+}, Al^{3+}, Cr^{3+}, Ni^{2+} and other impurities. Metals may be carried into the solution by parts treated in other processes or rinsing tanks, or they may be introduced by dissolution of the metals that compose the parts. The impurities could be present in concentrations from 10 to 25 gL^{-1}. These contaminants have undesirable effects on the plating solution including overvoltages, decrease in bath conductivity and in plating efficiency. As a result of bath contamination, the solutions frequently become spent and must be disposed of [4].

10.4 Conventional Treatment of Hexavalent Chromium Solutions

The major difficulty in the treatment of hexavalent chromium from wastewater solutions is the high concentration of chromium. Chromium in hexavalent form is soluble in acidic or alkaline pH. The most common and economical method for removing hexavalent chromium involves reducing it to its trivalent state and then causing its precipitation with an alkali. Ferrous ions, sulfur dioxide, sodium bisulfite or hydrazinium salts are used as reducing agents. The reduced trivalent chromium is then precipitated by addition of an alkali, such as lime or caustic soda, according to Eqs. (10.1) and (10.2).

$$4H_2CrO_4 + 6NaHSO_3 + 3H_2SO_4 \rightarrow 2Cr_2(SO)_4 + 3Na_2SO_4 + 1OH_2O \quad (10.1)$$

$$H_2Cr_2O_7 + 3NaHSO_3 + 3H_2SO_4 \rightarrow Cr_2(SO_4)_3 + 3NaSO_4 + 4H_2O \quad (10.2)$$

The amount of sludge produced in this process is very large, since for every 1 kg of chromium removed, 32 kg of sludge is generated, which is difficult to handle [5].

10.5 Electrodialysis for the Recovery of Water and Chromium Solutions

The purification of chromium solutions is interesting from both an environmental and economic perspective. This has motivated the development of efficient techniques for the separation and recovery of chromium ions. Membrane technology has become increasingly attractive for wastewater treatment and recycling. The main advantage of a membrane process is that concentration and separation are achieved without changing the physical state or using chemical products [6, 7].

In the chromium plating industry, the chromate ions may exist in the aqueous phase in different ionic forms ($HCrO_4^-$, CrO_4^{2-}, $Cr_2O_7^{2-}$, $HCr_2O_7^-$); the total amount of chromium and the pH dictate which particular chromium species will predominate in the aqueous phase. The CrO_4^{2-}anion prevails in basic or slightly acidic solutions, while the $Cr_2O_7^{2-}$ anion is dominant in acidic Cr^{6+} aqueous solutions. The metallic contaminants are found in cationic form, which are Al^{3+}, Cr^{3+}, Fe^{3+}, Ni^{2+}, Cu^{2+} [8].

In the last two decades the application of electrodialysis on chromium solutions has been studied. Several works for the reconcentration of solutions containing Cr^{6+} species have been published and patented [8, 9].

Some authors [10, 11] suggest the electrodialysis treatment technique because of the characteristics of the electroplating effluent, which presents a high level of dissolved ions and a low content of organic matter. The recovery of the first rinsing water after the chromium electrodeposition bath is performed by sending it to an electrodialysis system in order to concentrate and recover the chromium. The solution of the chromium recovered in the anode can be added to the original bath.

Other authors [12–14] suggest treating hexavalent chromium solutions with a membrane electrolysis system. Membrane electrolysis can regenerate process solutions through two primary mechanisms: (1) selective transfer of ions from the process solution, through the membrane, into an electrolyte solution and (2) regenerating oxidation states/ionic forms of key constituents in the process solution through electrochemical reactions at the electrode [15]. In one of the proposed systems [13], the rinsing water of the hexavalent chromium electroplating bath is pumped to an electrodialysis system, where the hexavalent chromium migrates towards the anode through an anionic membrane that retains contaminants of metallic ions (Cu, Fe, Ni) in the cathode compartment. The chloride ions present as contaminants are oxidized to chlorine at the anode.

Chen et al. [16] proposed a two-stage electrodialysis system to concentrate and purify chromate from electroplating wastewater with low pH using monovalent selective electrodialysis membranes. This process enabled the removal of monovalent impurities, such as chlorides.

Vallejo et al. [14] proposed a system using a cell with three compartments for the reconcentration of Cr^{6+}. The Cr^{6+} is transported through the anionic membrane (Ionac 3475-MA) and impurities (In and Ni) are transported through the cation exchange membrane (Nafion 450).

The treatment of acidic solutions of trivalent and hexavalent chromium by electrodialysis was studied by Dalla Costa. The authors used solutions of trivalent chromium sulfate and CrO_3 dissolved in HCl or H_2SO_4 employing Nafion 450, Selemion CMT and Selemion AMT membranes. They identified the reactions occurring at the electrodes and studied how these interfere in the treatment of chromium solutions with electrodialysis. The maximum yield in the process was reached when hexavalent chromium solutions were treated in the absence of sulfate ions [17].

Frenzel [9] studied the concentration of chromic acid rinse water by removing impurities such as Cr^{3+}, Fe^{3+}, and Ni^{2+}. A three compartment system was used where the central compartment contained the washing water and the other two compartments contained dilute H_2SO_4. The anionic membranes used were Ionac 3475 (Sybron Co.) and PC 100D (PCA GmbH), and the cationic membrane used was Nafion 324. In this system, the chromate anions migrate to the anode compartment through an anion membrane, forming chromic acid. This allows chromic acid to return to the deposition solution. The authors have shown that the concentration of Cr^{6+} in the anode compartment is dependent on the temperature and pH of the process. This study also shows that the removal of cationic contaminants is less efficient when compared to anodic contaminants [18].

Martí-Calatayud et al. [19] conducted galvanostatic experiments with an electrodialysis cell in order to study the passage of Ni^{2+} ions present in chromic acid solutions through a Nafion cation-exchange membrane. They studied the competing transport between H^+ and Ni^{2+} ions through the membranes.

The problem associated with the use of electrodialysis in the recovery of hexavalent chromium is usually linked to the poor stability of the anion-exchange membrane. So far, the development of resistant, highly conductive and selective membranes remains one of the main challenges for improving electrodialysis applications for the removal of heavy metals from industrial wastewater. Though polymeric homogeneous membranes possess good transport properties, they age rapidly when used with strong acidic and oxidizing mixtures, such as the effluents from plating industries [19].

Some studies were carried out to investigate aspects of the use of anion-exchange membranes in the separation and concentration of chromium (VI). Cengeloglu et al. [8] investigated some of the uses of anion-exchange membranes in the separation and concentration of chromium (VI). To accomplish this, three anion-exchange membranes (SB-6407, AFN and ACM) were tested and the effect on the salt solution was investigated. The authors observed that the chromium (VI) transport through anion exchange membranes depends on the nature of the electrolyte solution. The transport phenomenon seems to be closely connected to the structure of the membrane, such as cross-linking, water contents and proton leakage.

Korzenowski et al. [4] studied the purification of spent chromium baths contaminated with trivalent chromium, iron and aluminum. They verified that some membranes were more resistant to the oxidizing media than others. The resistance of the membranes was investigated as hexavalent chromium was passed through the cationic membrane to the catholyte. Chloride ions and/or sulfate ions hinder the passage of trivalent irons through the membranes. Total chromium interferes in the passage of iron and aluminum ions, much more for Fe^{3+} than Al^{3+} [4]. The degradation and incorporation of ions in different cation exchange membranes immersed in synthetic exhausted chromium bath was also evaluated. They verified that the Nafion 450 (Du Pont) membrane was the cationic membrane with the highest durability in a chrome bath and that the Selemion CMV, Selemion CMT and PC- SK (PCA) membranes showed very little durability, since the ion chrome was incorporated into all membranes.

Frenzel [18] evaluated the behavior of anion-exchange membranes in chromium baths. His work showed that the fumasep® FAP membrane seems to be a promising membrane for the recovery of chromic acid. It presented a superior performance than the other two commercial membranes studied (IONACMA3475 and PC 100 D). The main factors influencing the current efficiency of the Cr transport seem to be the chromic acid concentration in the anode compartment and the process temperature. Low chromic acid concentrations in the anode compartment and high temperatures result in significantly higher chromate transport [18].

10.6 Concluding Remarks

Although many studies have been conducted on the maintenance of chromium solutions using electrodialysis with anionic and cationic membranes, or membrane electrolysis with a cationic or an anionic membrane, the process is not yet applied on a large scale. New methods to recover hexavalent chromium baths are becoming increasingly important because of the amount of waste generated in the conventional process and because the costs for disposal are extremely high. The electrodialysis technique could be applied for the purification of hexavalent chromium baths, but it is important to modify the membranes to improve their resistance in oxidizing media. Among all the cationic membranes studied, Nafion®

proved to be the most resistant in the oxidizing medium. The fumasep® FAP anionic membrane seems to be a promising membrane for the recovery of chromic acid with the electrodialysis process.

References

1. Chaudhary AJ, Ganguli B, Grimes SM (2006) Concentrator cell methodology in the regeneration and recycle of chromium etching solutions using membrane technology. Chemosphere 62(5):841–846. doi: 10.1016/j.chemosphere.2005.04.074
2. Nriagu JO, Nieboer E (1998) Chromium in the natural and human environments. John Wiley & Sons, UK, p 571
3. Mandich N, Snyder DL (2010) Electrodeposition of chromium. In: Schlesinger Mordechay (ed) Modern Electroplating, 5th edn. Wiley, New York, p 205
4. Korzenowski C, Bresciani L, Rodrigues MAS et al (2008) Purification of spent chromium bath by membrane electrolysis. J Hazard Mat 152(3):960–967. doi:10.1016/j.jhazmat.2007.07.110
5. Reddithota D, Yerramilli A, Krupadarm RJ (2007) Electrocoagulation: a cleaner method for treatment Cr(VI) from electroplating industrial effluents. Indian J Chem Technol 14:140–245
6. Tay JH, Jeyaseelan S (1995) Membrane filtration for reuse of wastewater from beverage industry. Res Conserv Recycl 15(1):3340
7. Moura RCA, Bertuol DA, Ferreira CA et al (2012) Study of chromium removal by the electrodialysis of tannery and metal-finishing effluents. Int J Chem Eng 2012:1–7. doi:10.1155/2012/179312
8. Cengeloglu Y, Torb A, Kir E, Ersoz M (2003) Transport of hexavalent chromium through anion-exchange membranes. Desalination 154(3):239–246. doi:10.1016/S0011-9164(03)80039-5
9. Frenzel I (2005a) Waste minimization in chromium plating industry. Ph.D. thesis. University of Twente, Enschede, The Netherlands
10. Hartinger L (1994) Handbook of effluent treatment and recycling for the metal finishing industry. Carl Hanser, Munich
11. Pajunen P (1995) Hard chrome bath purification and recovery using ion-exchange. Met Finish 93(11):40–45. doi:10.1016/S0026-0576(05)80048-0
12. Knill EC, Chessin H (1986) Purification of hexavalent chromium plating baths. Plat Surf Finish 73 (8):26–32
13. Nonaka Y, Miyasaka T (1989) Regulating of bath concentration by electrodialysis of chromium plating solution. J Alum Finish Soc Kinki 138:1–7
14. Vallejo ME, Persin F, Innocent C et al (2000) Electrotransport of Cr(VI) through an anion exchange membrane. Sep Pur Tech 21:61–66
15. European Comission (2006) Integrated pollution prevention and control. Reference document on best available techniques for the surface treatment of metals and plastics. 2006. http://eippcb.jrc.es/reference/stm.html. Accessed 15 Jan 2013
16. Chen SS, Li CW, Hsua HD et al (2009) Concentration and purification of chromate from electroplating wastewater by two-stage electrodialysis processes. J Hazard Mat 161(2–3):1075–1080. doi: 10.1016/j.jhazmat. 2008.04.106
17. Dalla Costa RF, Rodrigues MAS, Ferreira JZ (1998) Transport of trivalent and hexavalent chromium through different ion selective membranes in acidic aqueous media. Sep Sci Technol 33(8):1135–1143. doi:10.1016/S0376-7388(02)00607-5
18. Frenzel I, Holdik H, Stamatialis DF et al (2005b) Chromic acid recovery by electro-electrodialysis I. Evaluation of anion-exchange membrane. J Membr Sci 261(1–2):49–67.doi: 10.1016/j.memsci.2005.03.031

19. Martí-Calatayud MC, García-Gabaldón M; Pérez-Herranz V(2012) Study of the effects of the applied current regime and the concentration of chromic acid on the transport of Ni^{2+} ions through nafion 117 membranes. J Membr Sci 392–393:137–149. doi: 10.1016/j.memsci.2011.12.012

Chapter 11
Electrodialysis Treatment of Metal-Cyanide Complexes

Marco Antônio Siqueira Rodrigues, Luciano Marder, Andréa Moura Bernardes and Jane Zoppas Ferreira

Abstract Metal-cyanide complexes have been used for many years as decorative and protective coatings on a variety of metal substrates. The most important method to protect iron and steel against corrosion is the application of zinc coatings. A number of processes have been developed to apply zinc coatings and the cyanide zinc plating process is still the most widely used. The method for treating wastewater containing metal-cyanide complexes is based on the transfer of the liquid phase to the solid phase. However, the solid residue produced in this treatment is dangerous and should be stored in an appropriate place, which represents a high cost. This treatment is being increasingly questioned because of the environmental consequences associated with the storage of toxic waste. Electrodialysis has been used as an alternative process for treating effluents in conventional electroplating industry, since it enables the recovery of water and the metal-cyanide complexes, minimizing or even avoiding the generation of galvanic sludge.

M. A. S. Rodrigues (✉)
Universidade Feevale, Novo Hamburgo–RS, Brazil
e-mail: Marcor@feevale.br

L. Marder
PPGSPI/UNISC. Santa Cruz do Sul,
Universidade de Santa Cruz do Sul, Santa Cruz do Sul–RS, Brazil
e-mail: lucmarder@yahoo.com.br

A. Moura Bernardes · J. Zoppas Ferreira
Programa de Pós Graduação em Engenharia de Minas, Metalúrgica e de Materiais (PPGE3 M), Universidade Federal do Rio Grande do Sul (UFRGS), Porto Alegre–RS, Brazil
e-mail: amb@ufrgs.br

J. Zoppas Ferreira
e-mail: Jane.zoppas@ufrgs.br

A. Moura Bernardes et al. (eds.), *Electrodialysis and Water Reuse*,
Topics in Mining, Metallurgy and Materials Engineering,
DOI: 10.1007/978-3-642-40249-4_11,

11.1 Introduction

The cyanide metal plating process is still the most widely used, despite the fact that its popularity has decreased significantly over the years across the world. Cyanide processes, however, are still the reference to which all other processes are compared, especially because the operating requirements are very stable for the coating process using metals and its alloys on a variety of metal substrates.

The most important method to protect iron and steel against corrosion is the use of metal coatings. Zinc is anodic with respect to iron and steel and provides protection when applied in thin layers. A number of processes have been developed for applying zinc coatings, including zinc electroplating using alkaline solutions, with or without cyanide, and acid solutions with chloride. The efficiency of zinc cyanide plating baths, however, is significantly better than that of zinc chloride baths.

The treatment of industrial effluents containing heavy metals is a major environmental problem that affects all countries in the world. The wastewater treatment method for metal electroplating effluents is based on the transfer of the liquid phase to the solid phase, including stages such as precipitation and filtration. The solid residue generated in this treatment is classified as hazardous and should therefore be stored in an appropriate place, which represents a high cost [5, 10, 14]. Besides the disadvantages of sludge generation, this treatment presents some deficiencies that are mainly related to the treatment of the wastewater, which contains metal-cyanide complexes. In this case, the metal cannot be separated by the conventional process. The metal-cyanide complexes should be destroyed before the metal precipitation stage. The most employed method for the treatment of wastewater containing cyanide is its destruction by oxidation to cyanate through an alkaline chlorination. However, this treatment can present some disadvantages, such as the low treatment efficiency of very stable metal-cyanide complexes, like those containing iron. Even though zinc cyanide complexes can be destroyed through alkaline chlorination, zinc ferrocyanide ($Zn_2Fe(CN)_6$) is formed in the presence of iron, a compound that is insoluble and will coprecipitate, causing contamination of the sludge with cyanide [28]. Another great inconvenience of this treatment is the possibility of the formation of organohalogenated compounds. These are highly toxic substances. Among the possible treatment alternatives, membrane technology offers many advantages, which are related with the general tendency towards resource conservations and energy management in the world. Because of their modularity, membrane techniques in general, and electromembrane techniques in particular, are very well suited for treating pollution at its source. The employment of electrochemical methods for the treatment of industrial wastewater has been increasing in the last years. Electrodialysis stands out among these methods because, in addition to not forming organohalogenated compounds, it has the great advantage of enabling the recovery and the reutilization of the cyanide, the metal-cyanide complexes and the water used in the industrial process [27], minimizing or even avoiding the generation of galvanic sludge.

11.2 The Transport of Metal Cyanide Complexes Through Ion Exchange Membranes

The behavior of metal ions in the presence of complexing agents is important, because the complex compounds change the fundamental properties of the metallic ions in solution, such as the charge, size and even the solubility of the ions. During the wastewater treatment by ED, the composition of the solutions is always changing because of the transport through the membrane. This change results in a modification of the species, which can even cause the precipitation of insoluble substances. Consequently, scaling problems can occur in the chambers of the diluate electrodialyzer [2].

The Nerst-Planck equation is used to describe, in dilute solutions, the transport of ionic species:

$$J_j = \left(-D \times \nabla C_j\right) + \left(\left(\frac{z_j \times F}{RT}\right) \times D_j \times C_j \times \nabla\phi\right) + \left(C_j \times v\right) \quad (11.1)$$

where, *Dj* is the diffusion coefficient of specie *j*, *Cj* is its concentration, *zj* is the valency, *F* is Faraday's constant, *R* is the universal gas constant, *T* is the absolute temperature, ϕ is the electrostatic potential and *v* is the velocity vector of the fluid. The three terms in the above equation represent the contributions of diffusion, migration and convection in the flow of species *j*, *Jj* in solution.

However, according to Akretche et al. [1, 2] the Nernst-Planck equations are difficult to be applied in solutions where complex species are present because of two main reasons: first, it is difficult to determine the diffusion coefficients of each species within the membrane. The second difficulty is associated with the wide variety of complex species present in the solutions. In addition, the distribution of these species varies during ED operation.

In the case of solutions containing complex species, the species transfer occurs from the ED-dilute (feed solution) to the ED-concentrate compartment according to the following steps:

(a) Transport of $[M(CN)_n]^{-m}$ or (CN^-) to the diffusion layer located in the solution of the ED-dilute compartment.
(b) Exchange of $[M(CN)_n]^{-m}$ or (CN^-) with the OH^- fixed ions inside the membrane according to the equilibrium presented in Eq. (11.2):

$$\left[M(CN)_n\right]^{-m}_{(1)} + m\, OH^-_m \Leftrightarrow \left[M(CN)_n\right]^{-m}_m + m\, OH^-_{(1)} \quad (11.2)$$

where (1) and *m* stand for the solution phase and the membrane, respectively.
(c) Exchange of $[M(CN)_n]^{-m}$ or (CN^-) with the OH^- ions located in the solution of the ED-concentrate compartment according to the equilibrium (Eq. 11.3):

$$\left[M(CN)_n\right]^{-m}_m + m\, OH^-_{(2)} \Leftrightarrow \left[M(CN)_n\right]^{-m}_{(2)} + m\, OH^-_m \quad (11.3)$$

where (2) stands for the solution in the concentrate compartment.

(d) Diffusion of the species $[M(CN)_n]^{-m}$ or (CN^-) in the concentrate compartment through the limiting layer.

Akretche et al. [2] state that the exchanges occur very rapidly and that the diffusion phase is the limiting step in the transfer kinetics. Gavach et al. [12, 13] state that two limiting effects should be taken into consideration in the application of electrodialysis to the treatment of solutions containing complex ions: the first is the competition for carrying current through the membrane between the complex ions and the free ions. The second is the possible occurrence of poisoning of the membrane by complexing agents and/or the complexed ions.

In this regard, Korngold et al. [15] investigated the influence of the hydrophobicity of complexing agents on the properties of membranes and observed that, in general, there is an increase in electrical resistance and a reduction in the selectivity of ion-selective membranes. These changes in the properties of membranes depend mainly on the nature (organic or inorganic) and the concentration of complexing agents in the solution. Sandeaux et al. [26] investigated the transport of organic ions through the anion exchange membrane and determined that the transport is influenced by the electrostatic interactions of organic ions with the functional groups or the polymer matrix of the membrane.

11.3 Solutions of Zinc Cyanide Complexes

The zinc oxide-hydroxide system is complicated by the number of forms that these compounds can take. The preparation method of zinc oxide affects the catalytic activity of the product. Inactive zinc oxide crystals usually exist as either short or long needles of a hexagonal structure. In these crystals, the zinc and oxygen atoms form a wurtzite type hexagonal structure. Under certain conditions, zinc atoms may be lodged in interstitial sites, giving rise to non-stoichiometric compounds. Each atom is surrounded by four oxygen atoms situated at the apices of a tetrahedron. The zinc hydroxide system is much more complex. The solid exists in five crystalline forms and also in an amorphous form. At ordinary temperatures, the only stable form is the &-hydroxide and all the other forms are converted into this form, although the transformations are not simple [29].

Zinc can generate different complex species, depending on the concentration, pH and type of ions present in the solution. The distribution of the metal among the several soluble complexes can be calculated from the equilibrium constants [4].

Zinc in an alkaline medium can generate different complex species, depending on the concentration and on the kind of ions present in the solution. In an alkaline medium the species present in the solution are: Zn^{2+}, $Zn(OH)^+$, $Zn(OH)_2$, $Zn(OH)_3^-$, $Zn(OH)_4^{2-}$, $Zn(OH)_n^{(n-2)-}$. The distribution of zinc species in alkaline solutions is reported in the literature. According to Dirkse [8], zinc is mainly present as $[Zn(OH)_4]^{2-}$ in hydroxyl concentrations exceeding 2 M. In the presence

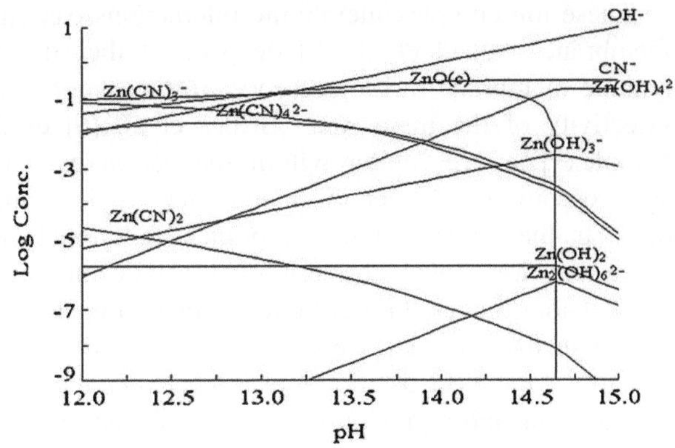

Fig. 11.1 Species distribution diagram for Zn^{2+} in aqueous solutions

of cyanide there are also other species: $ZnCN^{+}$, $Zn(CN)_2$, $Zn(CN)^{3-}$, $Zn(CN)_4^{2-}$, $Zn(CN)_n^{(n-2)-}$ [21].

The zinc in an alkaline medium, and in the presence of cyanide, can form complexes with hydroxyl ions and with cyanide ions, with an equilibrium between compounds with cyanide and hydroxyl, as can be seen in the Eq. (11.4):

$$[Zn\ (CN)_4]^{-2} + 4OH^- \Leftrightarrow [Zn\ (OH)_4]^{-2} + 4CN^- \qquad (11.4)$$

In this reaction, an increase in the hydroxyl concentration displaces the equilibrium towards the formation of tetra-hydroxyzincate. On the other hand, an increase in the concentration of cyanide ions promotes the formation of tetra-cyano-zincate. The other ionic species present in solution are free cyanide, hydroxide and sodium ions. If the total cyanide concentration is sufficiently high, all metals in solution will be present as an anionic complex [21].

Figure 11.1 shows the species present in the solutions of zinc in an alkaline medium in the presence of cyanide.

11.4 The Transport of Zinc Cyanide Complexes Through Ion Exchange Membranes

The transport of complexes of zinc ions through ion selective membranes has been investigated by several researchers. Aouad et al. [3] investigated the properties of the anionic membrane in equilibrium with aqueous solutions containing zinc chloride complexes, linking the variations in the electric resistance and in the amount of water in the membrane with the presence of zinc complex ions. The decrease in the amount of water inside the membranes occurs because voluminous ions such as $[ZnCl_3]^-$ are balanced to fixed places of the membrane, which reduces the available space for water within the membrane.

These ion complex/membrane interactions can cause profound changes in the membrane. Aouad et al. [3] determined the sodium ion transport through an anionic membrane in the presence of zinc complexes, indicating the loss of the selectivity of the membrane. Bribes et al. [6] established the presence of the complex $[ZnCl_3]^-$ anion within the cation exchange membrane with Raman spectroscopy, even after the membrane was kept immersed in water for several days. In the ion-selective membranes, functional groups linked to the polymer chain are associated through electrostatic interactions with mobile solvated counter-ions that are brought into the membrane along with the water. The entry of ions with the same charge of the fixed functional groups (co-ions) is prevented because of the electrostatic repulsion between these two species. This attraction by counter-ions and repulsion by co-ions of functional groups is the mechanism that explains the selectivity of the ion-selective membranes. The transport of positive ions through the anionic membrane indicates that the complex ions are acting as counter ions of the active sites of the membranes.

The transport of the cyano-complexes of zinc and copper was studied by Chiapello and Gal [7]. The authors observed that the resistance of the anionic membrane increases with rising concentrations of cyano-complex ions in solution. In addition, the presence of complex ions in solution does not cause changes in the cationic membrane resistance, since the metal complex ions do not penetrate this membrane.Rodrigues et al. [22] investigated the influence of the presence of zinc complex species on the selectivity of the anionic AMV membrane to sodium ions (Na^+). The results show a high percentage extraction of sodium monovalent cations through the anionic AMV membrane in the presence of complex zinc/cyanide. The occurrence of this transport indicates a loss of permselectivity of the AMV anionic membrane in the presence of zinc-cyanide complexes, showing a strong interaction between the zinc-cyanide complexes and the active sites of the membrane. This interaction is so significant that the selectivity of the AMV membrane is substantially altered, which was verified by the high sodium monovalent cation transport through the anionic membrane. Since no sodium transport across the AMV anionic membrane was observed in the presence of complex zinc-hydroxyl and of cyanide and hydroxyl ions, there is an indication that the interactions that occur between these ions and the membrane are not strong enough to cause changes in the selectivity of the membrane. The determination of the permselectivity coefficient of zinc-cyanide complexes relative to zinc-hydroxyl complexes was determined as: $T_{\mathrm{Zn(OH)}_4^{-2}}^{\mathrm{Zn(CN)}_4^{-2}} = 12,32$. This result demonstrates that the interaction with the AMV anionic membrane of complex ions of zinc-cyanide is greater than the interaction of complex ions of zinc-hydroxyl.

In the same study, the transport of zinc complex ions through an anion exchange membrane for the treatment of solutions with and without cyanide was evaluated. The ionic transport was analyzed as a function of the applied current density and of the cyanide and hydroxyl concentrations. Table 11.1 presents the chemical composition of the solutions used in this work [22].

Table 11.1 Chemical composition of the solutions [22]

Solution	ZnO	NaCN	NaOH
A	15 $g{\cdot}L^{-1}$	–	90 $g{\cdot}L^{-1}$
B	15 $g{\cdot}L^{-1}$	15 $g{\cdot}L^{-1}$	70 $g{\cdot}L^{-1}$
C	15 $g{\cdot}L^{-1}$	30 $g{\cdot}L^{-1}$	–

Figure 11.2a, b, c presents the percent extraction of zinc complex ions that permeate through the anionic AMV membrane as a function of the varying current density, tests durations and solutions applied (Table 11.1).

In Fig. 11.2a, it is possible to observe that the transport of zinc complex ions through the AMV membrane is related to both parameters (time and current density). The results show that an increase in the applied current density causes ion transports through the membrane to increase. Nevertheless, Fig. 11.2b shows that there is an increase in ion transport with current densities varying from 5.2 to 29 $mA{\cdot}cm^{-2}$, but there is a decrease in permeability with a current density of 40.7 $mA{\cdot}cm^{-2}$. This phenomenon is probably associated with the come out of a limiting current density to the ionic transport. Similar behavior was observed by Dibenedetto and Lighfoot [9], who identified a current limit in their experiments, from which point the transport of ions is controlled by diffusion in the solution.

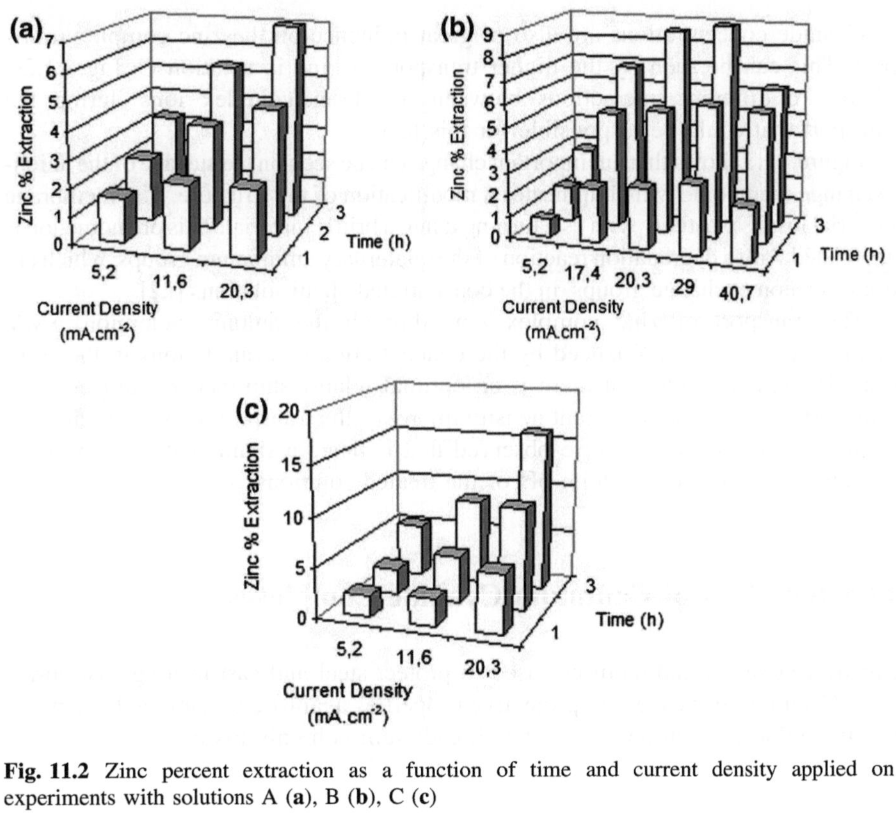

Fig. 11.2 Zinc percent extraction as a function of time and current density applied on experiments with solutions A (**a**), B (**b**), C (**c**)

Fig. 11.3 Micrographs of the superficial morphology of the AMV membrane used in the experiments

Cyanide concentrations are also of great influence on the zinc complex transport. This can be seen by the higher transport of zinc in solution C, Fig. 11.2c. There are different interactions between membrane/complex ions during the transport that could be responsible for this fact.

Figure 11.3 shows that an important change can be seen on the surface of the anion-exchange membrane, which indicates a modification of the structure. The membrane material lost its plasticity with use, turning it into a brittle material. This phenomenon is associated with a degradation reaction of the quaternary ammonium groups, which are used as anion exchange groups, in the concentrated alkali solutions [22].

The transport of zinc complex ions through the anionic Selemion AMV membrane is greatly influenced by the concentration of cyanide ions in the solution. The results show that there is an optimal relationship that maximizes ionic transport. An increase in current density improves the transport. However, changes in the anionic membrane were observed that indicate a significant change in the membrane because of the high pH of the treated solutions.

11.5 Solutions of Cadmium Cyanide Complexes

Electrodeposits of cadmium are used to protect steel and cast iron against corrosion. Cadmium has excellent protective properties against corrosion and has many useful engineering properties, although cadmium is highly toxic.

Cadmium can be electroplated from solutions with different compositions. Despite the fact that there are two types of cadmium baths—those that use acid electrolytes and those using alkaline based cyanide electrolytes [23]—cadmium coatings are produced almost exclusively from alkaline cyanide-based baths, mainly because of their high penetrating power and the high quality of deposits obtained. The alkaline cyanide-based baths are normally prepared by dissolving cadmium oxide or cadmium cyanide in a solution of sodium cyanide [19, 25].

Cadmium and cyanide, in different proportions, lead to the formation of different cadmium cyanide complexes in alkaline media. According to Prytz and Osterud [24], in the interval Cd:CN 1:4 and Cd:CN 1:20, the $Cd(CN)_4^{2-}$ complex should prevail. Flengas [11] suggests that with excess cyanide all the cadmium is complexed in the $Cd(CN)_4^{2-}$ form. Ortega and collaborators [20] suggest that $Cd(CN)_4^{2-}$ is formed when there is a great excess of cyanide in a solution with pH higher than 10. Koivula and co-workers [16] demonstrate that for a Cd:CN 1:10 system, 95 % of cadmium occurs in the $Cd(CN)_4^{2-}$ form.

In a typical cadmium alkaline-based cyanide bath (CdO 30 $g{\cdot}L^{-1}$, NaCN 105 $g{\cdot}L^{-1}$, NaOH 18.8 $g{\cdot}L^{-1}$), the predominant cadmium cyanide complex formed is $Cd(CN)_4^{2-}$. The other ionic species present in the solution are free cyanide (CN^-), hydroxide (OH^-) and sodium (Na^+) ions. In an electrodialysis cell, sodium ions are transported through the cation-exchange membrane, whereas metal complexes, free cyanide and hydroxide ions are transported through the anion-exchange membrane [7].

11.6 The Transport of Cadmium Cyanide Complexes Through Ion Exchange Membranes

The treatment of a cadmium solution by electrodialysis was studied by Marder et al. [17]. The removal of the $Cd(CN)_4^{2-}$ and CN^- ions from a 1.14 g L^{-1} CdO, 3.99 g L^{-1} NaCN and 0.71 g L^{-1} NaOH solution, using a five-compartment electrodialysis cell, shows that more CN^- ions are removed than $Cd(CN)_4^{2-}$ ions. This is probably associated to the ionic radii, since the cadmium cyanide complex is more voluminous than the cyanide ion. The highest removal rate for $Cd(CN)_4^{2-}$ and CN^- ions in this system was 86 % and 95 %, respectively. However a cadmium compound precipitation on the cation-exchange membrane placed between the cathodic concentrate and the diluted compartments was observed. This may be a consequence of the distribution of ionic species, which varies during the ED operation. Such precipitation is not observed in removal rates up to 23 % $Cd(CN)_4^{2-}$ and 43 % CN^-.

Figure 11.4a shows the micrograph obtained by SEM (Scanning Electron Microscopy) for the cation-exchange membrane used up to 86 % and 95 % $Cd(CN)_4^{2-}$ and CN^- ions removal, respectively. Figure 11.4b shows the corresponding Energy Dispersion Spectroscopy (EDS) spectra of the membrane [17].

The cadmium compound precipitate reduces the process efficiency and causes irreversible damage to the membrane. The electrodialysis process should be stopped before the precipitation occurs. However, when the electrodialysis process is stopped before the precipitation of the cadmium compound, the removal of $Cd(CN)_4^{2-}$ and CN^- is small. Alternatives should be used in order to make the electrodialysis process operate to satisfaction. One alternative tested is the

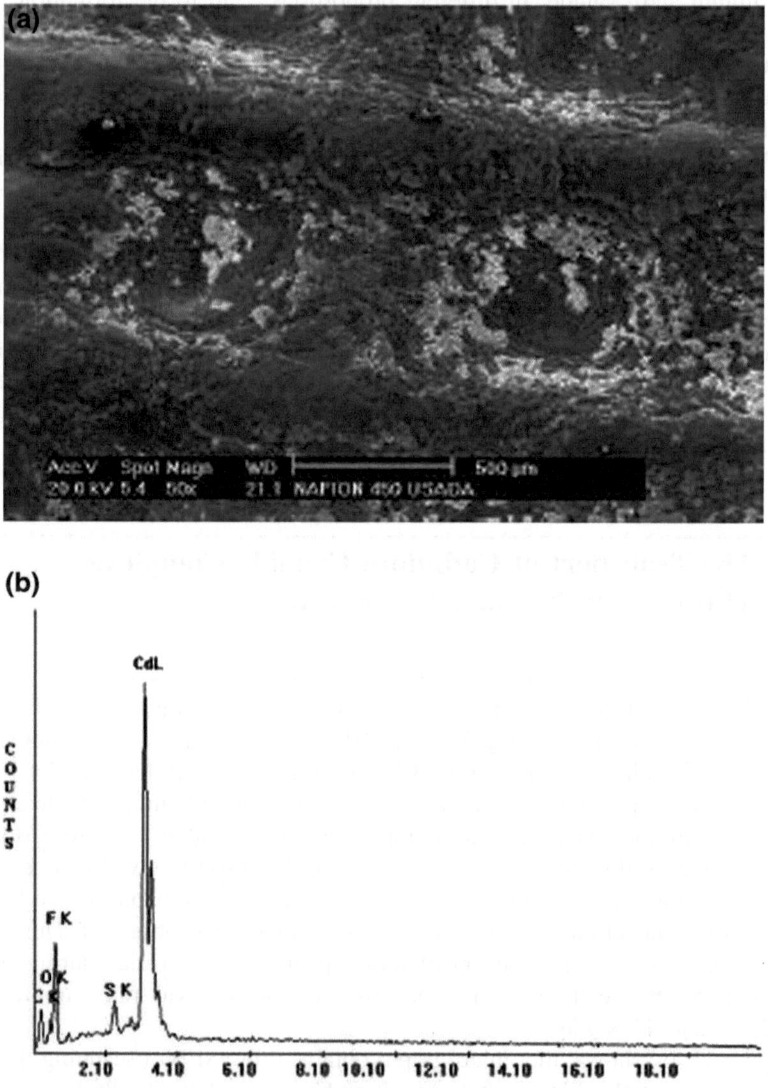

Fig. 11.4 **a** Used cation-exchange membrane (NAFION 450) micrograph and **b** respective EDS spectra [17]

Table 11.2 Percent extraction (%) of $Cd(CN)_4^{2-}$ and CN^- during electrodialysis experiments

Time (min)	Percent extraction (%)	
	$Cd(CN)4^{2-}$	CN^-
120	20.1	44.1
240	20.1	42.4
360	19.2	41.1
480	18.7	38.2
600	19.0	39.4

Solution: 1925 mg l^{-1} $Cd(CN)4^{2-}$, 1175 mg l^{-1} CN^- , 710 mg l^{-1} NaOH. Ion-exchange membranes Nafion 450, Selemion AMV. Current density: $15 mA cm^{-2}$

successive change of the feeding solution before the precipitation. Experiments were done (Table 11.2) changing the solution of the diluate cell compartment for each 120 min of electrodialysis.

This procedure was repeated four times, resulting in a total of 600 min of electrodialysis. With these experiments, it is possible to obtain a solution that is 56 % more concentrated in cadmium and 250 % more concentrated in cyanide in the anodic concentrate cell compartment than the initial feeding solution of the diluate cell compartment. This demonstrates that it is possible to obtain a more concentrated solution of cadmium and cyanide in the other cell compartments without the precipitation of the cadmium compound on the cation-exchange membrane. The obtained solution can then be reused in the cadmium electroplating bath. The $Cd(CN)_4^{2-}$ and CN^- percent extraction presents a small decrease during the 600 min of electrodialysis [18].

11.7 Influence of the Presence of Other Metallic Ions (Copper, Iron, Chromium)

A chemical characterization of 15 samples of industrial rinse water of a cadmium electroplating process showed that this wastewater is characterized by the presence of other metallic ions than cadmium (chromium, aluminum, silver, iron, nickel, copper, lead and zinc), all of them in lower concentrations than cadmium. The data obtained in an evaluation of the influence of copper, iron and chromium on the removal/recovery of $Cd(CN)_4^{2-}$ and CN^- ions, revealed that the transport of these three metallic ions occurs in the same direction as $Cd(CN)_4^{2-}$ and CN^- ions, i.e. these ions occur predominantly in anionic form. These ions also form charged coordination complexes between the metallic ions and the CN^- and OH^- ligands. This represents a negative factor in the recovery of cadmium and cyanide, since it represents a contamination of the electroplating bath. In addition, the removal of $Cd(CN)_4^{2-}$ ions is also reduced in the presence of the metallic ions, which should be associated with the competition between the metallic ions and $Cd(CN)_4^{2-}$. Nevertheless, water recovery is still possible with this method, since a good removal rate was obtained for all metal cyanide complexes [18].

11.8 Concluding Remarks

Electrodialysis can be an alternative for the treatment of effluents from electroplating processes containing metal cyanide complexes. The behavior of metal in the presence of hydroxyl and cyanide ions changes fundamental properties of the metal ions in solution, such as the charge, size and even the solubility of the ions. The recovery of metal complexes with electrodialysis is associated with the OH^- and CN^- concentrations in the solutions. There is an optimum molar ratio between the concentration of cyanide, hydroxyl and metal ions in solution. For values above or below this ratio the recovery of metal decreases. The results show that there is an optimal relationship that maximizes ionic transport. However, the anionic membrane can be attacked by strong alkaline (above pH 12) media. This means that solutions of metals must be treated with a pH below this threshold.

References

1. Akretche DE, Kerdjoudj H, Gherrou A (1997) Electrodialysis of solutions obtained by elution of cyanide complexes from anionic exchange resin by means of acidic thioure. Hydrometall 46(3):287–301. doi:10.1016/S0304-386X(97)00026-1
2. Akretche DE, Kerdjoudl H (2000) Donnan dialysis of copper, gold and silver cyanides with various anion exchange membranes. Talanta 51(2):281–289. doi:10.1016/S0039-9140(99)00261-1
3. Aouad F, Lindheimer A, Gavach C (1997) Transport properties of electrodialyssis membranes in the presence of Zn^{+2} complexes with Cl^-. J Membr Sci 123:207–223
4. Blackie MS, Gold V (1959) The stability of the $Zn(CN)_4^{2-}$ ion. J Chem Soc 4:3932–3934
5. Brazil (2004) Associação Brasileira de Normas Técnicas. ABNT NBR 10004:2004. Solid Wastes—Classification
6. Bribes JL, Huguet P, Chaouki M (1996) Ramam spectroscopy investigation and improved knowledge on industrial cation-exchange membranes involved in electrodialysis process. J Mol Struct 379:219–226
7. Chiapello JM, Gal JY (1992) Recovery by electrodialysis of cyanide electroplating rinse waters. J Membr Sci 68(3):283–291. doi:10.1016/0376-7388(92)85029-I
8. Dirkse TP (1954) The nature of the zinc-containing ion in strongly alkaline solutions. J Electrochem Soc 101(6):328–331
9. Dibenedetto AT, Lighfoot EN (1962) The separation of ions with permselective membranes. J. AIChE 8(1):79–86. doi:10.1002/aic.690080120
10. EPA530-R-96-008 (1996) International waste minimization approaches and policies to metal plating. Office of Solid Waste
11. Flengas SN (1955) Polarographic and potentiometric titration of the cadmium cyanide complexes. Trans Faraday Soc 51:62. doi:10.1039/TF9555100062
12. Gavach C, Elmidaoui A, Cherif AT (1993) Separation of Ag^+, Zn^{2+}, and Cu^+ ions by electrodialysis with monovalent cation specific membrane and EDTA. J Membr Sci 76(1):39–49. doi:10.1016/0376-7388(93)87003-T
13. Gavach C, Kerdjoudj H, Akretche DE (1993) Electrodialysis of copper-thiourea solutions. Hydrometall 34:231–238. doi:10.1016/0304-386X(93)90037-E
14. Hartinger L (1994) Handbook of effluent treatment and recycling for the Metal Finishing Industry, 2nd edn. Carl Hanser, Munich

15. Korngold E, De Korosy F, Rahay R et al (1970) Fouling of anion selective membranes in electrodialysis. Desalination 8:195–220
16. Koivula R, Lehto J, Pajo L et al (2000) Purification of metal plating rinsewaters with chelating ion exchangers. Hydrometall 56(1):93–108. doi:10.1016/S0304-386X(00)00077-3
17. Marder L, Sulzbach GO, Bernardes AM et al (2003) Removal of cadmium and cyanide from aqueous solutions through electrodialysis. J Braz Chem Soc 14(4):610–615. doi:10.1590/S0103-50532003000400018
18. Marder L, Bernardes AM, Ferreira JZ (2004) Cadmium electroplating wastewater treatment usinga laboratory-scale electrodialysis system. Sep Purif Technol 37(3):247–255. doi:10.1016/j.seppur.2003.10.011
19. Morrow H (2000) Cadmium electroplating. Met Finish 98(1):210–214. doi:10.1016/S0026-0576(00)80327-X
20. Ortega E, Pérez-Herranz V, Guiñón JL et al (2001) Effect of cadmium content on cyanide transport through anionic membranes. Proceedings of the international water conference-IWC, Porto
21. Rodrigues MAS, Korzenovski C, Gondran E et al (2006) Evaluation of changes on ion selective membranes in contact with zinc-cyanide complexes. J Membr Sci 279(1–2):140–147. doi:10.1016/j.memsci.2005.11.045
22. Rodrigues MAS, Zoppas JF, Amado FDR et al (2008) Transport of zinc-complexes through anion exchange membrane. Desalination 227(1–3):241–252. doi:10.1016/j.desal.2007.07.018
23. Panossian Z (1993) Corrosão e proteção contra corrosão em equipamentos e estruturas metálicas. IPT, São Paulo, p 636
24. Prytz M, Osterud TH (1952) Polarographic investigations of dilute solutions of cadmium cyanide complexes. Acta Chem Scand 6:1534–1544
25. Safranek WH (1974) The properties of electrodeposited metals and alloys. American Elsevier Publishing Company Inc, New York, p 517
26. Sandeaux R, Sandeaux J, Delimi R et al (1995) Properties of an anion exchange membrane in contact with aqueous solutions of sodium chloride and sodium benzenecarboxylate or benzenesulfonate. J Membr Sci 103(1–2):83–94. doi:10.1016/0376-7388(94)00310-U
27. Schmieder H, Galla U, Juttner K (2000) Electrochemical approaches to environmental problems in the process industry. Electrochim Acta 45(15–16):2575–2594. doi:10.1016/S0013-4686(00)00339-X
28. Tanihara K, Yasuda S, Tamai K (1983) Treatment of wastewater from cyanide zincx electroplating. Met Finish 81(1):5–53
29. Thirsk HR (1974) Electrochemistry, vol 4. Alden-Mowbray LTd, Oxford

Chapter 12
Electrodialysis Treatment of Nickel Wastewater

Tatiane Benvenuti, Marco Antônio Siqueira Rodrigues, Andréa Moura Bernardes and Jane Zoppas Ferreira

Abstract The galvanic processes are one of the main activities contributing to metal discharges into the environment. A large volume of wastewater is generated that contains a high load of salts and metals and it must be treated to recover the chemicals and water and save resources. Nickel is a toxic metal and causes various health problems. According to environmental regulations across the world, nickel concentrations in effluents must be controlled on an acceptable level before the discharge into the environment. The removal of metals by conventional treatment (chemical precipitation) not only does not result in a final effluent with a nickel concentration below the acceptable limit, but it also generates a large volume of galvanic sludge, a hazardous waste material. Several treatment processes have been suggested for the removal of nickel from rinse water, such as electrochemical techniques. This chapter presents the application of electrodialysis (ED) as an alternative that can contribute to comply with legal environmental standards and enable the recovery and reuse of water and chemicals in the nickel electroplating process, helping to minimize the environmental impact associated with the water consumption and generation of waste in the galvanic industry.

T. Benvenuti (✉) · A. Moura Bernardes · J. Zoppas Ferreira
Programa de Pós-Graduação em Engenharia de Minas, Metalúrgica e de Materiais (PPGE3M), Universidade Federal do Rio Grande do Sul (UFRGS), Porto Alegre–RS, Brazil
e-mail: tati.eng.biobio@gmail.com

A. Moura Bernardes
e-mail: amb@ufrgs.br

J. Zoppas Ferreira
e-mail: jane.zoppas@ufrgs.br

M. A. S. Rodrigues
Programa de Pós Graduação em Qualidade Ambiental (PPGQA), Universidade Feevale, Novo Hamburgo-RS, Brazil
e-mail: marcor@feevale.br

A. Moura Bernardes et al. (eds.), *Electrodialysis and Water Reuse*,
Topics in Mining, Metallurgy and Materials Engineering,
DOI: 10.1007/978-3-642-40249-4_12,

12.1 Introduction

Electroplating or galvanic processes are based on metal plating baths and rinse water. They generate a large volume of liquid effluent with harmful metal concentrations of Ni, Cr, Cu, Zn, etc. Water consumption in these processes is very substantial, and the chemicals used are expensive and come from non-renewable sources. The electroplating wastewater should therefore be treated to permit the reuse of water and the recovery of chemicals, avoiding their discharge into the environment.

The traditional treatment methods are not efficient, because they generate galvanic sludge that is considered a hazardous waste [8] and which can cause environmental contamination. Nickel can cause some acute and chronic effects on human health, such as gastric and skin diseases [18]. Nickel compounds, such as nickel sulphide, are suspected to cause cancer [8, 29]. Electrodialysis is becoming a good alternative in comparison to the traditional wastewater treatment methods.

12.2 The Use of Nickel in Electroplating Processes

Nickel electroplating is one of the most versatile surface-finishing processes available. It has a broad spectrum of end uses that encompass decorative, engineering, and electroforming applications [9]. Nickel is widely used in galvanic industry [29], where it is applied on steel, zinc alloys, copper and its alloys, and chemically metalized plastic [11]. Nickel electroplating is an old process: the first experiences date back to 1840. The first nickel baths used available nickel salts, such as double nickel salts, mineral acid salts, and acetates. Weston recognized the importance of boric acid. Later, Bancroft recognized the importance of nickel chloride. Finally, in 1916, Watts combined nickel sulfate, nickel chloride, and boric acid to produce the modern Watts-type nickel bath in use today [21].

Electroplated nickel can be mat or bright, according to the bath used. Nickel plating has the function of leveling the imperfections, which allows a deposition with an excellent aspect. In industrial environments, the product can be coated with chrome after the nickel plating, avoiding the tarnish caused by sulfur and increasing the corrosion resistance.

Decorative coatings are obtained by electroplating with special solutions containing organic addition agents. Generally, the property sought in engineering end uses is corrosion resistance, but wear resistance, solderability, magnetic properties and other characteristics may be relevant for specific applications. To obtain these distinct coating characteristics, different nickel bath compositions are applied.

12.3 Conventional Nickel Wastewater Treatment

Like other galvanic process, nickel plating is based on metal plating baths and rinse water and generates effluents with a metal concentration that varies according to the applied process [4]. The nickel bath is periodically reinforced. The rinse water is discharged and can contain up to 1,000 mgNi/L [8].

The treatment of electroplating effluents in the galvanic industries usually applies operations of sedimentation and filtration of hydroxides, sulfides or carbonates. These methods are attractive because they enable a high degree of metal removal and do not require expensive equipment. However, the major disadvantage of this conventional physicochemical process is the large volume of galvanic sludge that is produced. This is a hazardous waste that needs to be discharged in appropriate industrial waste disposal centers, which can be very expensive [15].

Furthermore, the addition of NaOH (to get pH 10 for the precipitation of nickel compounds) and other chemicals may be necessary to ensure the quality of the treated effluent. It is very difficult to comply with the limits of legislation for nickel and other pollutants in the treated water with this method. Additional processes are therefore necessary to treat this effluent. One example is the coagulation-floculation method, which uses iron salts and polymeric floculants: these chemicals confer high salinity to the water, which in most cases eliminates the possibility of water reuse. The treated effluent, once it meets the standards set out in legislation, is therefore discharged in public sewers, the soil or in rivers [13].

From an environmental protection and resource conservation perspective, the effective recycling and reuse of the metal wastewater is strongly recommended. Closed–loop systems or so-called effluent free technologies, should be developed. Some clean technologies are already in use: adsorption [18], evaporation [13], ion exchange resins [12], membrane processes like nanofiltration, ultrafiltration and reverse osmosis [20], and electrochemical processes like electrolysis [19] and electrodialysis, which is the subject of this book.

12.4 Electrodialysis for Water and Nickel Recovery

Electrodialysis can be applied to the recover water and nickel in a metal surface treatment process. Nickel plating includes several rinsing processes. The nickel concentration in the effluent from the first rinsing stage is high. The ED process is designed to collect nickel ions in the first rinsing stage and return them to the electroplating bath in order to increase the recovery ratio of nickel and decrease the nickel content in the waste water from the final rinsing stage [25]. Figure 12.1 represents an ED process operating in the galvanic industry.

When compared to electrolysis, where the metal is recovered and deposited on the electrode, ED has the advantage of recovering the metal solution which can then be added to the bath in operation. The deposited metal, on the other hand,

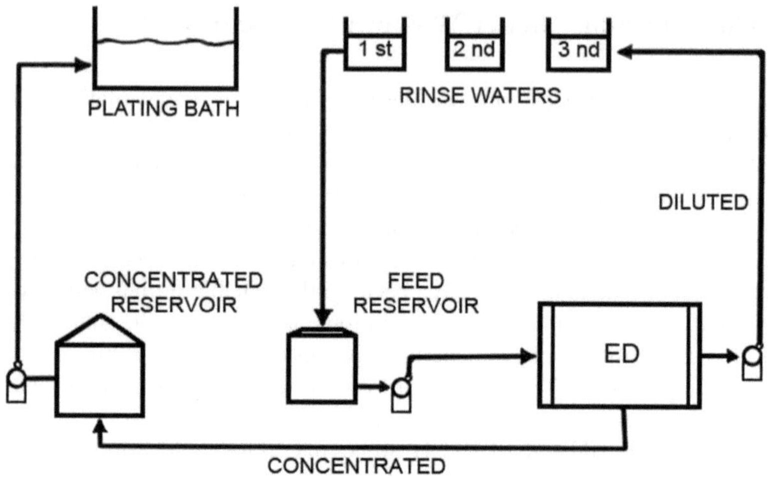

Fig. 12.1 Electrodialysis in an electroplating process, to recover water, metal and salts

must be dissolved to be returned to the bath. One of the limitations of ED as an electroplating solution is the fact that is not possible to recover the additive components of the baths, mostly organic compounds, if they are not ionized [23]. Many studies have been conducted with real and synthetic wastewater in order to apply this technique to remove various concentrations of nickel of the first rinsing water to recover the components of the electrodeposition bath and to reuse as much water as possible.

Earlier works have demonstrated the successful application of electrodialysis in nickel solutions and date back to 1972 [27]. This studies have encouraged industrial applications and research to increase the treatment efficiency. Tison and Mikhail [26] evaluated ED using a low current to recover and recycle nickel (salts) from dilute wastewaters. In a three compartment cell, using nickel metal as electrodes and two membranes (configuration cathode-cationic membrane-anionic membrane-anode), a diluted solution from a Watts nickel bath (4,000 ppmNi) was fed to the central compartment. The others compartments were filled with a concentrated nickel bath. Results showed that nickel with typical rinse water concentrations can be transferred electrically across commercially available membranes, and directly into a concentrated Watt's nickel plating bath (72,000 ppm). Recovery with approximately 90 % current efficiency was possible with current densities as low as 3.0 $mAcm^{-2}$. Using lower current densities, the efficiency dropped, depending on the nickel diffusion through the membranes (without current application, nickel was transported by diffusion through the cationic membrane at a rate of $7.1 \times 10^{-5} gh^{-1} cm^{-2}$). Nickel and co-transported water were routinely recycled as a relatively concentrated solution which would not cause dilution [26].

Li and co-workers [14] assayed the recovery of spent electroless nickel plating baths by electrodialysis. The purpose of recovering the spent bath is to effectively remove the harmful ions (phosphite, sulfate, sodium), while retaining useful ions, such as nickel, hypophosphite ions, organic acids. Because the useful life of plating bath is prolonged through recovery treatment, the cost of electroless nickel plating is decreased. One of the advantages of ED treatment is that nickel removal is minimal when compared to other ions, since the formation of nickel-organic acid complexes reduces or even neutralizes the metal charge. They have a larger ionic radius than Ni^{2+}, limiting their passage through the membranes.

Crotty and Bailey [7] evaluated ED to lengthen the electroless nickel bath life using four ED designs. The standard off-line batch design involves taking a process solution off line, after the bath has been used. This method was the least efficient of the four designs, since the solution to be treated was heavily contaminated. The frequent batch design treats the electroless nickel process solution every night during down times, and it is more efficient, because the bath is purified frequently in small increments. The continuous high-temperature design is an on-line process that is used at the same time that the electroless nickel process solution is being used for production. This process design is more efficient than the other designs. The continuous low-temperature design is another on-line process that is used as the electroless nickel bath production solution is being used. This design is expected to have similar advantages as the continuous high temperature design.

Bernardes and co-workers [4] evaluated current density and pH of synthetic effluents based on the Watts bath composition, since this is the most commonly used bath, in treatments with a two compartment ED cell. Tests at lower pH indicated an ion transport competition between H^+ and Ni^{2+}, reducing the nickel extraction, independent of the applied current. Agitation was necessary during ED for effluents with a pH of 5 to avoid the precipitation of nickel.

Bouhindel and Rumeau [5, 6] investigated membrane fouling and the different electrodialytic behavior and properties of components in nickel plating rinse waters during ED treatment. These researches indicated different behavior of pH at the solutions during the ED experiments to the treatment of $NiCl_2$ and $NiSO_4$. Another observation indicated the cumulative effect of boric acid on the membrane, causing fouling, which is one of the problems of ED for nickel effluent treatment.

Spoor and co-workers [22] evaluated tests in laboratory and in a pilot plant of a galvanic industry during three months. Two tests were performed during this period, under the same operational parameters, to treat rinse water containing 5 $mg{\cdot}L^{-1}$ Ni^{2+}. A three compartments cell, with Nafion® 117 cationic membranes and Ti/Pt electrodes was used. The cathode compartment was filled with 1 M H_2SO_4 and the anode compartment, with 1 M HCl. The effluent was a synthetic effluent prepared from a dilution of a Watts nickel bath in operation in the galvanic plant, which was fed to the central ED compartment, with a cation exchange resin. Their tests showed the effect of pH on the hybrid ED-ion exchange resin process. If $pH < 2$, the Ni^{2+} fraction adsorbed by the resin decreases and the removal rate also reduced. However, the reduction of pH inhibits the formation of $Ni(OH)_2$ in the ion-exchange compartment. The treatment was successful on the pilot scale, since the initial concentration was reduced from 5 ppm to <20 ppb. The solution flow was the critical factor for the process efficiency. The tests were performed continuously, without formation of hydroxide or other solid compounds, indicating that the treatment cell can operate without human intervention.

Tzanetakis and his workgroup [28] evaluated Nickel and Cobalt extraction from sulfate solutions using electrodialysis to compare the efficiency of two cation exchange membranes: the perfluorosulfonic Nafion® 117 (commercial) and a new sulfonated PVDF membrane, under similar operating conditions. The membranes were used as either flat structures or as corrugated structures. The corrugated structure is obtained when flat membranes are submitted to a controlled mechanic pressure and temperature, which enables the formation of small channels (2 mm × 1 mm, w × h). There is no detectable damage to the membrane material. This structure provides larger surface area for ion transport and thus an increased ionic flow. In the three compartment cell, separated by a cation exchange membrane and an anion exchange membrane, the electrolyte solution with 0.2 M $NiSO_4$ (11.72 $mg{\cdot}L^{-1}$) fed the anode compartment, the central compartment contained 0.01 M H_2SO_4 while the catholyte contained 1.0 M H_2SO_4 to increase the conductivity of the cell. When the flat membranes were used, applying a current density of 40 $mA{\cdot}cm^{-2}$ and 2$L{\cdot}min^{-1}$ as solution flow, a current efficiency of 69 % was observed. As for the nickel extraction efficiency the corrugated membranes were observed to be 3–5 % more efficient than the flat structure. The effect of current density on the NI^{2+} transport number indicated a better use of current under 20 $mA{\cdot}cm^{-2}$, reducing the concentration polarization effects and NI^{2+}/H^{+} competition for both membrane structures. The authors also evaluated a mixed metal solution containing nickel and Co sulfate (2,000 ppm each metal) using the new sulfonated PVDF cation and anion exchange membranes. EDTA was added to allow the selective extraction of Co at the catholyte. A $[Ni\text{-}EDTA]^{-2}$ complex was preferentially formed and the complexed nickel ions were retained in the central compartment, since the bigger radius of this complex inhibits its transport through the anion exchange membrane [28].

Taghdirian and coworkers [24] also surveyed electrodeionization for the selective separation of NI^{2+}/CO^{2+} ions from dilute aqueous solutions. The separation of cobalt and nickel ions was done using ED in the presence of EDTA, exploiting the greater stability of the EDTA complex with nickel. The authors used a three compartment ED cell, where the middle compartment was separated from the other two by two cation Exchange membranes (CR67 MKIII, Ionics, USA). The middle compartment was filled with strongly acid cation exchange resins (Purolite, C100E), initially in the H^{+} form. A platinum coated titanium mesh and stainless steel electrodes were used as anode and cathode, respectively. Sulfuric acid solutions (0.05 M) were pumped into the electrodes compartments. The nickel remained in solution as a negatively charged complex because the Donnan exclusion effect prevents the entrance of this complex into the gel phase of the resins and membrane.

Marder et al. [16, 17] studied nickel transport and the effects of sulfate and chloride species, the presence of boric acid, ammonium chloride and other metallic ions on chronopotenciometric techniques. The limiting current density was determined for $NiSO_4$ and $NiCl_2$ solutions containing 1,400 ppm of nickel and was equal to 2 $mA \cdot cm^{-2}$ and 2.86 $mA \cdot cm^{-2}$. Although the limiting current density was higher, the transport number was lower in the chloride medium. The nickel transport number through the cation exchange membrane in the sulfate medium was similar for 0.01 and 0.05 $mol \cdot L^{-1}$ concentrations, but the transport number was reduced when the H^+ concentration increased, since H^+ competes with Ni^{2+} for the current transference. It was not verified any interference of H_3BO_3 during the ED for $NiCl_2$ solutions, the transport number of nickel did not change and the presence of boric acid minimized the water dissociation and pH fluctuation in the ED system.

Dzyako and Belyakov [10] studied Nickel removal in diluted $NiSO_4$ solutions by electrodeionization. The central compartment in a three compartment ED cell, separated by a cation exchange membrane and an anion exchange membrane, was filled with different cationNi^{2+} exchange resins. Using Pt as electrodes, different H_2SO_4 concentrations were evaluated as electrode solution. The diffusion coefficient through the resin increased when the pH of the effluent was reduced, increasing the nickel concentration in the catholyte. More acid cathode solutions also reduced $Ni(OH)_2$ precipitation on the cation exchange membrane. For the removal of 0.35 mol of nickel from a 1 m^3 solution containing 1 mol of $NiSO_4$, the energy consumption was 208 Wh.

Benvenuti et al. [3] studied the effect of pH, electroplating bath additives and membrane stack configuration on ED in the treatment of synthetic and actua rinse water from nickel electroplating. 16 cm^2 membranes and $Ti/Ti_{0.7}Ru_{0.3}O_2$ electrodes were used. The cells were compared under similar operating conditions. The stack configurations tested are shown in the Fig. 12.2.

Synthetic solutions containing $NiCl_2$, $NiSO_4$ and H_3BO_3 containing 2 $gNi \cdot L^{-1}$ were evaluated. In addition, the actual rinse water contained organic compounds and higher nickel concentrations (around 4.7 $g \cdot L^{-1}$). The transport of nickel ions through cationic membranes (C) occurred as expected, but the transport of nickel

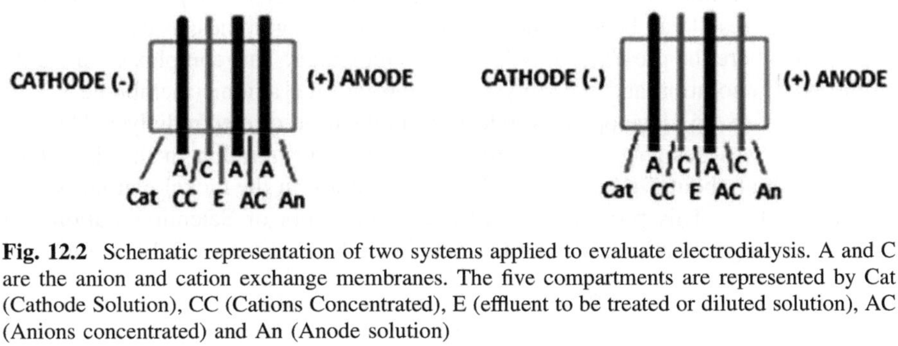

Fig. 12.2 Schematic representation of two systems applied to evaluate electrodialysis. A and C are the anion and cation exchange membranes. The five compartments are represented by Cat (Cathode Solution), CC (Cations Concentrated), E (effluent to be treated or diluted solution), AC (Anions concentrated) and An (Anode solution)

ions through one or two anion exchange membranes (A) to the anodic compartments, in very low concentrations, was also observed in both stack configurations and independent of the absence or presence of the electroplating bath additives. In both experiments it was possible to verify that the anode solution became green because of the presence of nickel. The cathode solutions became gray only in the ED experiments with the actual effluent, which could indicate a reduction of organic compounds in the cathode. The pH was adjusted during the experiments when the pH of the effluent achieved values around 6–7, reducing it to 3. Nevertheless a large amount of $Ni(OH)_2$ precipitated on the cation exchange membrane and on the anion exchange membrane in the cathode compartment. When the effluent is acidified, the nickel transport rate decreases because of the competition between H^+ and Ni^{2+}. The nickel transport rate is a similar parameter to transport number and indicates a relationship between nickel concentration (Ni), treated solution volume (V), membrane area (S), ED time (t) and current applied (I), according to the Eq. (12.1).

$$\tau = \frac{\left([Ni_0] - [Ni_f]\right)(mgL^{-1}) \times V(L)}{S(cm^{-2}) \times t(h) \times I(A)} \tag{12.1}$$

At the beginning of experiments, the actual effluent presented 2.6 $g{\cdot}L^{-1}$ of COD, and at the end, the value decreased to 1.6 gL^{-1} of COD. When the others compartments were studied, the highest values for COD were observed only in the anionic concentrate solution, indicating the negative charge of the organic molecules.

Later studies by [2] supported that the nickel transport through the anion exchange membranes occurred with both Chinese HDX-200 and Italian Ionac® MA-3475 membranes, regardless of the presence of organic additives, and that the nickel transport rate during ED for nickel extraction was similar in both cation exchange membranes and effluents tested. The ED results for $NiSO_4$ solution was compared to $NiCl_2$ solutions and different synthetic effluents, containing only one organic plating additive, to investigate the production of negatively charged nickel with organic compounds. All results indicated nickel transport through the anion exchange membranes. This result means that the additives, although organic, did not restrict the application of the membrane technique in the studied wastewater treatment. Only in the ED for $NiCl_2$ solution, no nickel transport was detected through the anionic membrane when the pH was corrected using HCl instead of H_2SO_4. This result can be explained by Hydra-Medusa graphics, as can be seen at Fig. 12.3, where the presence of $Ni(SO_4)_2^{-2}$ is indicated. This complex occurs only in the $NiSO_4$ solution and probably passes through the anionic membrane.

Asahi Glass Co. developed an industrial application of electrodialysis [1]. The ED system was designed to maintain the nickel concentration in the first stage rinsing bath at about 5 $g{\cdot}L^{-1}$, when the concentration in the nickel plating bath is about 84 $g{\cdot}L^{-1}$. This process uses 40 membrane pairs of Selemion Cation and Anion exchange membranes, with 0.336 m^2 of effective area and 1.0 $A{\cdot}dm^{-2}$ of

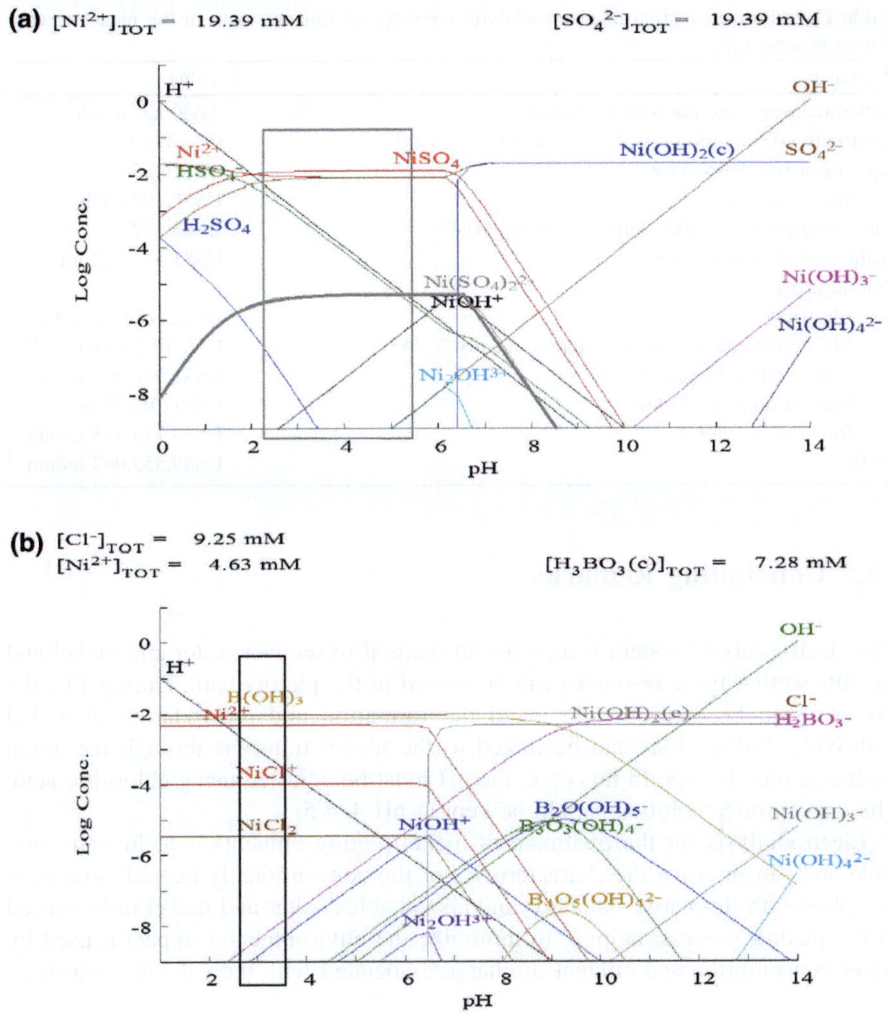

Fig. 12.3 Ionic species in the $NiSO_4$ **a** and $NiCl_2$ **b** solutions. The pH range during ED was 2.3–5.5

current density. In this process the recovery ratio exceeds 90 % and a diluted solution is perfectly recycled to the first stage rinsing bath. Current efficiency and the required electric power for ED were greater than 90 % and 2 $kWh.kg^{-1}$ Ni ion, respectively. The cost estimation that was performed by the company is presented in Table 12.1 [25].

Table 12.1 Cost estimation of electrodialytic recovery of rinsing waste in the nickel electroplating process [25]

Parameter	Value
Nickel recovery rate (as $NiSO_4 \cdot 6H_2O$)	3,640 kg month^{-1}
Electricity consumption (as $NiSO_4 \cdot 6H_2O$)	0.7 kWh kg^{-1}
Equipment instalation cost	US$160,500
Electricity unit cost	US$0.107 kWh^{-1}
Purchising price of nickel salt (as $NiSO_4 \cdot 6H_2O$)	US$3.852 kg-1
Profit of Nickel salt recovery	US$1,3327.92 month^{-1}
Running cost	
Electricity	US$259.154 month^{-1}
Maintence and consumable item (annually 3 %)	US$401.25 month^{-1}
Amortization (7 years)	US$1,910.699 month^{-1}
Interest (annually 9 %)	US$1,203.75 month^{-1}
Running cost total	US$3,774.853 month^{-1}
Profit	US$9,553.067 month^{-1}

12.5 Concluding Remarks

The electrodialysis system is an efficient method to recover water and nickel and its salts so that these resources can be reused in the plating bath. During ED, the pH needs to be controlled to avoid the formation and precipitation of nickel hydroxide. Sulfate ions can be linked to the nickel transport through the anion exchange membranes. In this case, the pH must be adjusted using chloridric acid. The concentrated solution should be kept at pH 4–4.5.

Electrodialysis for the treatment of nickel plating effluents is technically feasible and can improve the characteristics of the conventionally treated effluent. It complies with discharge standards and also enables water and nickel to be reused in the plating process, helping to minimize the environmental impact caused by water consumption and effluent discharge associated with the galvanic industry.

References

1. AGC ASAHI GLASS (2011) Available at: http://www.agc.com/english/. Accessed in May 2013
2. Benvenuti T (2012) Avaliação da eletrodiálise no tratamento de efluentes de processos de eletrodeposição de níquel. (Evaluation of electrodialysis in the treatment of nickel electroplating wastewaters). Master Thesis. PPGE3M–UFRGS. Porto Alegre, Brazil
3. Benvenuti T, Bordignon GL, Fensterseifer Jr G et al (2012) Influence of the electrodialysis stack configuration on the treatment of nickel electroplating effluent. In: Libro de Resúmenes—VIII Congreso Ibero-Americano en Ciencia y Tecnología de Membrana—CITEM, Universidad Nacional de Salta, p 283–284. Argentina, 24–27 April 2012

4. Bernardes AM, Dalla Costa RF, Fallavena VLV et al (2000) Electrochemistry as a clean technology for the treatment of effluents: The application of electrodialysis. Met Finish 98(11):52–58, 114. doi:10.1016/S0026-0576(00)83558-8
5. Bouhidel KE, Rumeau M (2000) Comparison of the electrodialytic properties on $NiSO_4$ and $NiCl_2$: Influence of the salt nature in electrodialysis. Desalination 132(1–3):195–197. doi:10.1016/S0011-9164(00)00149-1
6. Bouhidel K E, Rumeau M (2004) Ion-exchange membrane fouling by boric acid in the electrodialysis of nickel electroplating rinsing waters: generalization of our results. Desalination 167:301–310. doi: 10.1016/j.desal. 2004.06.139
7. Crotty DE, Bailey DE (2002) Electrodialysis of electroless nickel process solutions: Continuous versus batch treatment designs. Met Finish 100(11–12):30–31,33–39. doi: 10.1016/S0026-0576(02)80933-3
8. Dermentzis K (2010) Removal of nickel from electroplating rinse waters using electrostatic shielding electrodialysis/electrodeionization. J Hazard Mater 173(1–3):647–652. doi:10.1016/j.jhazmat.2009.08.133
9. DiBari GA (2000) Nickel Plating. Metal Finishing. International Nickel Inc., Saddle Brook, N.J. (98) 1:270–288. doi: 10.1016/S0026-0576(00)80334-7
10. Dzyazko YS, Belyakov VN (2004) Purification of a diluted nickel solution containing nickel by a process combining ion exchange and electrodialysis. Desalination 162:179–189. doi:10.1016/S0011-9164(04)00041-4
11. Groshart EC (1997) Preparation of Basis Metals for Plating. In: Michael Murphy (ed) Metal Finishing, (95), 1A:189–199. doi:10.1016/S0026-0576(99)80018-X
12. Juang RS, Kao HC, Chen W (2006) Column removal of Ni(II) from synthetic electroplating wastewater using a strong-acid resin. Sep Purif Technol 49(1):36–42. doi:10.1016/j.seppur.2005.08.003
13. Kurniawan TA, Chan GYS, Lo W-H et al (2006) Physico-chemical treatment techniques for wastewater laden with heavy metals. Chem Eng J 118(1–2):83–98. doi:10.1016/j.cej.2006.01.015
14. Li CL, Zhao HX, Tsuru T et al (1999) Recovery of spent electroless nickel plating bath by electrodialysis. J Membr Sci 157(2):241–249. doi:10.1016/S0376-7388(98)00381-0
15. Lin X, Burns RC, Lawrance GA (1998) Effect of electrolyte composition, and of added iron (III) in the presence of selected organic complexing agents on Nickel (II) precipitation by lime. Water Res 32(12):3637–3645. doi:10.1016/S0043-1354(98)00131-6
16. Marder L, Navarro EMO, Pérez-Herrans V et al (2006) Evaluation of transition metals transport properties through a cation-exchange membrane by chronopotentiometry. J Membr Sci 284(1–2):267–275. doi:10.1016/j.memsci.2006.07.039
17. Marder L, Navarro EMO, Pérez-Herranz V et al (2009) Chronopotentiometric study on the effect of boric acid in the nickel transport properties through a cation-exchange membrane. Desalination 249: 348–352. doi:10.1016/j.desal.2009.06.040
18. Nabarlatz D, Celis J, Bonelli P et al (2012) Batch and dynamic sorption of Ni(II) ions by activated carbon based on a native lignocellulosic precursor. J Environ Manag 97:109–115. doi:10.1016/j.jenvman.2011.11.008
19. Orhan G, Arslan C, Bombach H et al (2002) Nickel recovery from the rinse waters of plating baths. Hydrometal 65(1):1–8. doi: 10.1016/S0304-386X(02) 00038-5
20. Quin JJ, Wai MN, Htun M et al (2004) A pilot study for reclamation of a combined rinse from a nickelplating operation using a dual-membrane UF/RO process. Desalination 161:155–167
21. Schario M (2007) Troubleshooting decorative nickel plating solutions (Part I of III installments): Any experimentation involving nickel concentration must take into account several variables, namely the temperature, agitation, and the nickel-chloride mix. Met Finish 105(4):34–36. doi: 10.1016/S0026-0576(07)80584-8
22. Spoor PB, Grabovska L, Koene L et al (2009) Pilot scale deionization of a galvanic nickel solution using a hybrid ion-exchange/electrodialysis system. Chem Eng J 89:193–202

23. Strathmann H (1995) Electrodialysis and related processes. In: Noble RD and Stern SA (eds.) Membrane Separations Technology: Principles and Applications, Elsevier Science, p 213–281. doi:10.1016/S0927-5193(06) 80008-2
24. Taghdirian HR, Moheb A, Mehdipourghazi M (2010) Selective separation of Ni(II)/Co(II) ions from dilute aqueous solutions using continuous electrodeionization in the presence of EDTA. J Memb Sci 362(1–2):68–75. doi:10.1016/j.memsci.2010.06.023
25. Tanaka Y (2007) Ion Exchange Membranes: Fundamentals and Applications, vol 12, Elsevier, pp 1–531
26. Tison RP, Mikhail YM (1982) Electrodialysis performance when recycling dilute nickel solutions directly into a plating bath. J Memb Sci 11(2):147–156. doi:10.1016/S0376-7388(00)81397-6
27. Trivedi DS, Prober R (1972) On the feasibility of recovering nickel from plating wastes by electrodialysis. Ion Exch Membr 1:37
28. Tzanetakis N, Taama WM, Scott K et al (2003) Comparative performance of ion exchange membranes for electrodialysis of nickel and cobalt. Sep Purif Technol 30(2):113–127. doi:10.1016/S1383-5866(02)00139-9
29. Zhao M, Duncan JR (1998) Removal and Recovery of Nickel from aqueous solution and electroplating rinse effluent using *Azolla filiculoides*. Process Biochem 33(3):249–255. doi:10.1016/S0032-9592(97)00062-9

MIX
Papier aus verantwortungsvollen Quellen
Paper from responsible sources
FSC® C105338

If you have any concerns about our products,
you can contact us on
ProductSafety@springernature.com

In case Publisher is established outside the EU,
the EU authorized representative is:
Springer Nature Customer Service Center GmbH
Europaplatz 3, 69115 Heidelberg, Germany

Printed by Libri Plureos GmbH
in Hamburg, Germany